India Studies in Business and Economics

The Indian economy is considered to be one of the fastest growing economies of the world with India amongst the most important G-20 economies. Ever since the Indian economy made its presence felt on the global platform, the research community is now even more interested in studying and analyzing what India has to offer. This series aims to bring forth the latest studies and research about India from the areas of economics, business, and management science. The titles featured in this series will present rigorous empirical research, often accompanied by policy recommendations, evoke and evaluate various aspects of the economy and the business and management landscape in India, with a special focus on India's relationship with the world in terms of business and trade.

More information about this series at http://www.springer.com/series/11234

Barun Deb Pal • Vijay P. Ojha • Sanjib Pohit
Joyashree Roy

GHG Emissions and Economic Growth

A Computable General Equilibrium Model Based Analysis for India

 Springer

Barun Deb Pal
Centre for Economic Studies and Policy
Institute for Social and Economic Change
  (ISEC)
Bangalore
Karnataka
India

Vijay P . Ojha
Institute of Management Technology
Ghaziabad
Uttar Pradesh
India

Sanjib Pohit
CSIR–National Institute for Science
  Technology and Development Studies
  (CSIR-NISTADS)
New Delhi
India

Joyashree Roy
Department of Economics
Jadavpur University
Kolkata
West Bengal
India

ISSN 2198-0012   ISSN 2198-0020 (electronic)
ISBN 978-81-322-1942-2   ISBN 978-81-322-1943-9 (eBook)
DOI 10.1007/978-81-322-1943-9
Springer New Delhi Heidelberg New York Dordrecht London

Library of Congress Control Number: 2014944812

Preface

We decided to get this book published to help the larger audience make sense of the discussion on climate change and its implications for economic growth in India. Issues linking climate change and economic growth are now at the centre of discussions regarding climate friendly development strategies which are increasingly becoming a necessity rather than an option for late-industrialising countries like India. This book contributes to this discussion by systematically analysing the relationships between economic growth and GHG emissions in India with explicit reference to all major economic sectors. Although the science of the impact of climate change on earth and its inhabitants is moving rapidly in the direction of certainty and precision, lack of clarity on how emerging economies can manage their developmental imperatives in the face of pressing carbon constraints with judicious policy interventions persists. Whether a global carbon price will help incentivising developmental actors to choose low-carbon growth strategies is still an open and debatable issue. Many suggest carbon tax at the border or nationally implemented under sovereign national fiscal regimes. Suggestions for benchmarking-monitoring-reporting-verification of all activities by best practices are also on the table. So, it is quite a complex issue for anybody to try to resolve without detailed knowledge and information on pros and cons of each of the alternative interventions and institutional arrangements suggested. Moreover, the impact of climate change is all pervading. It is not confined to one single activity or sector but extends to each and every economic activity and sector and to people of all socioeconomic groups.

Our joint effort in this area began in 2006 when India's Ministry of Environment and Forest, gave a small research grant to National Council of Applied Economic Research (NCAER) and Jadavpur University (with the former and latter being then the institutions of affiliation for the first three authors and the fourth author respectively), to provide them with knowledge and analytical support on India's GHG emissions profile. The necessity and inevitability of pooling multiple expertises to get this empirical investigation accomplished brought all of us together with the lead author of the book also availing the opportunity to find an interesting and relevant research topic for his Ph.D. work which he eventually completed by virtue of hard labour and patience to our great pleasure.

This book estimates latest Social Accounting Matrix (SAM) for India. It provides a very important database describing the complete circular flow of income and input-output transactions among the sectors of the economy. Striking novelty of the book lies in the fact that for the first time to the best of our knowledge, a SAM for Indian economy has been prepared with environmental indicators and detailed methodology is also presented in the book. The environmental social accounting matrix (ESAM) based analysis has been included in the book to show direct and indirect linkage between economic growth and GHG emissions.

The work we present here goes beyond SAM and applies computable general equilibrium (CGE) modelling to conduct climate change policy analysis and simulations. The analysis is an important contribution in the current debate around carbon tax and its possible impacts on macroeconomic growth. Knowledge sharing by Dr. Pradipto Ghosh on CGE modelling as applied to climate change issues needs special mention.

During the course of this detailed work we received help from a number of excellent people in various forms. Administrative support from Prof. Binay Kumar Pattanayak, Director ISEC, Director General of NCAER, Dr. Shashanka Bhide, senior research councilor, NCAER, Dr. Anushree Sinha, Senior Fellow, NCAER, Mr. N. J. Sebastian, Former Secretary and Librarian, NCAER; and Dr. Nandita Bhattacharyya of Faculty of Arts, Staff members of Department of Economics-Jadavpur University, made the progress of the work smooth. Comments and academic advice received from Prof. Pradeep Biswas, CSIR-NISTADS, Prof. Rajashree Majumder, University of Burdwan, Prof. M. R. Naryana, ISEC, Prof. K. V. Raju, ISEC, Prof Meenakshi Rajeev, ISEC, and Mrs. B. P. Vani, ISEC needs special acknowledgement. Study material collected by NCAER library staff, from Dr. Maniparna Shyam Roy, and Dr. Duke Ghose of Global Change Programme of Jadavpur University were immensely useful. Lastly, we thank for never ending family support for each one of us.

What has driven us and kept us together until we finished this book is the genuine wish to advance the knowledge on the subject and deep concern for saving our planet and ourselves from extinction if global warming remains unchecked.

Last but not least, the results expressed in this book are those of the authors and are not attributable to the institute/organization to which they belong.

Barun Deb Pal
Vijay P. Ojha
Sanjib Pohit
Joyashree Roy

Contents

Contributors

Barun Deb Pal Centre for Economic Studies and Policy, Institute for Social and Economic Change (ISEC), Bangalore, Karnataka, India

Vijay P. Ojha Institute of Management Technology, Ghaziabad, Uttar Pradesh, India

Sanjib Pohit CSIR–National Institute for Science Technology and Development Studies (CSIR-NISTADS), New Delhi, India

Joyashree Roy Department of Economics, Jadavpur University, Kolkata, West Bengal, India

About the Authors

Dr. Barun Deb Pal is currently an Assistant Professor at the Institute of Social and Economic Change, Bangalore. He has been working in the field of SAM based CGE modelling for the last 7 years. He has also worked on climate-smart agriculture and land-use planning models for South Asia as a key researcher at the International Food Policy Research Institute (IFPRI). He has a long professional affiliation with the National Council of Applied Economic Research (NCAER), working on climate change CGE modelling. He has published papers in journals of international repute with one of his publication being "A Social Accounting Matrix for India", published in 2012 in Economic Systems Research (Taylor and Francis). His current areas of research interest include various developmental issues and their linkages with climate change, low carbon agriculture, Infrastructure, and public utility pricing.

Dr. Vijay P. Ojha is a Professor of Economics at the Institute of Management Technology, Ghaziabad, India. A Computable General Equilibrium modeler (CGE) by training, he employed the CGE modelling technique to analyze the trade-offs among carbon emissions, economic growth and poverty reduction in India while tenuring as a post-doctoral Commonwealth Fellow at the Environment Department, University of York, United Kingdom in 2004. Ever since, he has been working on applications of CGE modelling in climate change and other areas like trade, human capital and inclusive growth. He has authored many published reports and academic journal papers His most recent publications are in the Journal of Policy Modeling (Elsevier) and Environment and Development Economics (Cambridge University Press).

Dr. Sanjib Pohit Professor AcSIR and Senior Principal Scientist, is presently working at CSIR- National Institute of Science, Technology & Development Studies (CSIR-NISTADS). He was educated at the Indian Statistical Institute. Previously, he held research positions (Senior Fellow/Chief Economist) at the National Council for Applied Economic Research, New Delhi and the Indian Statistical Institute and was also a visiting scholar at the University of Michigan (Ann Arbor, USA) and the Conference Board of Canada. He is an experienced modeler in the area of trade and environment with 20 years of modelling experience. He has he worked in the area of institutional economics, transport economics, input-output models, FDI, informal trade, the automobile industry, and South Asian integration. He has co-authored 7 books, and has published more than 100 articles in journals/books. He has presented his research at seminars and conferences in different parts of the world—Japan, Canada, USA, India, Bangladesh, Switzerland, the Netherlands, Singapore, and Finland.

Dr. Joyashree Roy, an ICSSR National Fellow, is currently a Professor of Economics at Jadavpur University, Kolkata, India. Additionally, she coordinates the Global Change Programme and also directs the Ryoichi Sasakawa Young Leaders Fellowship Fund Project at Jadavpur University. In 2007 she was on the Nobel Peace Prize winning panel –IPCC (Intergovernmental panel on climate change). She has been involved in preparation of Stern Review Report, Global Energy Assessment and many other national and global reports. She has published more than 70 peer-reviewed articles in journals of national and international repute. Her research interests are in resource and environmental economics, particularly in the field of energy demand modelling, economic assessment of application of non-renewable and renewable resources for urban and rural problems, economic and social dimensions of climate change, water resource management and multidisciplinary research on sustainable development, sustainability transition, ecosystem services in the context of coastal ecosystem.

Abbreviations

AEEI	Autonomous energy efficiency improvement
AGRIM	Agriculture, growth and redistribution of income model
ANFI	Aggregate non-factor input
ASI	Annual Survey of Industries
CAB	Current account balance
CDM	Clean development management
CE	Compensation to employees
CES	Constant elasticity of substitution
CET	Constant elasticity of transformation
CGE	Computable general equilibrium
COP	Conference of the parties
CO_2EQ	Carbon equivalent
ESAM	Environmental social accounting matrix
FDI	Foreign direct investment
GAMS	General algebraic modelling systems
GCMs	General circulation models
GDP	Gross domestic product
GEF	Global Environment Facility
GVA	Gross value added
HYV	High-yielding variety
IO	Input output
KgOE	Kg of oil equivalent
LES	Linear expenditure system
LFPR	Labour force participation rate
MBI	Market-based instruments
MoEF	Ministry of Environment and Forests
NAMEA	National Accounting Matrix for Environmental Accounting
NAS	National accounts statistics
NATCOM	National communication
NCAER	National Council of Applied Economic Research
NHPCL	National Hydro Power Corporation Limited
NPCIL	Nuclear Power Corporation of India Limited

NSSO	National Sample Survey Organization
NVA	Net value added
OS	Operating surplus
PCE	Per capita emission
PFCE	Private final consumption expenditure
PPP	Purchasing power parity
PPM	Parts per million
PPB	Parts per billion
SAM	Social accounting matrix
SDA	Structural decomposition analysis
SGM	Second generation model
TERI	The Energy and Resource Institute
TFPG	Total factor productivity growth
TL	Translog
UNFCCC	United Nations Framework Convention on Climate Change
US$	US dollar

List of Figures

List of Tables

Appendix

Chapter 1
Economic Growth and Greenhouse Gas (GHG) Emissions: Policy Perspective from Past Indian Studies

Understanding the impact of climate change on the economy's performance has become an important issue for all the countries: developed, developing, or emerging. Some see this problem as more acute in case of developing countries which are on high growth trajectories (Aggarwal and Narain 1991). The emissions of greenhouse gases (GHGs) have increased over time (Das et al. 2007; Sharma et al. 2006; Intergovernmental Panel on Climate Change (IPCC) 2010; Stern 2007; GEA 2012). The IPCC in its fourth assessment report mentioned that the changes in atmospheric concentration of GHGs and aerosols, land cover, and solar radiation alter the energy balance of the climate system and become drivers of climate change. According to this report, the annual carbon dioxide (CO_2) concentration growth rate has been larger during the last 10 years (1995–2005 average 1.9 parts per million (ppm) per year) than it has been since the beginning of continuous direct atmospheric measurements (1960–2005 average 1.4 ppm per year). The global atmospheric concentration of methane has increased from a preindustrial value of about 715 parts per billion (ppb) to 1732 ppb in the early 1990s, and is 1774 ppb in 2005. The global atmospheric nitrous oxide concentration increased from a preindustrial value of about 270 ppb to 319 ppb in 2005. The precipitation has become spatially variable and the intensity and frequency of extreme events have increased. The sea level also has risen at an average annual rate of 1–2 mm during this period. However, the continued increase in concentration of GHGs in the atmosphere is likely to lead to climate change resulting in large changes in ecosystems, leading to possible catastrophic disruptions of livelihoods, economic activity, living conditions, and human health (IPCC 2010).

Historically, the industrialized countries have been the primary contributors to GHG emissions. Only 25% of the global population living in Annex I countries emit more than 70% of the total global CO_2 emissions and consume 75–80% of many of the other resources of the world (Parikh et al. 1991). But, because of their high population and economic growth rates, the fossil fuel use led CO_2 emissions from developing countries are likely to soon match or exceed those from the industrialized countries (Sathaye et al. 2006). Therefore, if the responsibility for emissions increase in the past lies largely with the industrialized world, then the late-industrializing countries are likely to be the source of an increasing proportion of future increase of GHGs.

B. D. Pal et al., *GHG Emissions and Economic Growth*,
India Studies in Business and Economics, DOI 10.1007/978-81-322-1943-9_1,
© Springer India 2015

1.1   Climate Change and Economic Growth: Global Context

It is now widely accepted in climate change literature that the global economy is highly vulnerable to global warming, with sectors most directly and heavily impacted being agriculture, coastal resources, energy, forestry, tourism, and water (Pearce et al. 1996 and Bach et al. 2001). But the economies of some countries are more vulnerable to climate change than those of others because of varying share of these sectors in economic growth (Blackman and Harrington 2000). Late-industrializing countries in general have a larger share of their economic activity coming from agriculture and forestry (Darwin et al. 1995). They also tend to be in the lower latitudes where the impacts of these sectors will be most severe. The low latitudes tend to be too hot for reasonably profitable agricultural activities and any further warming will further reduce agricultural productivity and thereby agricultural profitability here. Up to 80 % of the damages from climate change may be concentrated in low latitude countries (Mendelsohn 2006).

It is generally agreed that the countries in Africa will experience declining yields in the long run. For example, agricultural production in Guinea-Bissau where agricultural sector adds value of up to 62 % of gross domestic product (GDP) is estimated to contribute only 32.7 % of GDP (without carbon fertilization) by 2080. The impacts on development and food security, as well as on nutrition, will be enormous (Mendelsohn 2006).

1.2   Climate Change and Economic Growth: Indian Context

Indian economy is highly vulnerable to global warming caused by GHG emissions. The Indian Network for Climate Change Assessment report (INCCA 2010) indicates that the average annual surface air temperature in India is increasing by 0.40℃ with not much variation in absolute rainfall. The sea level has increased at a rate of 1.06–1.25 mm/year during the last four decades across the coastal India. The same report has predicted that the temperature in India will be increased by 2–40℃ by 2050s. The climate sensitive sectors such as agriculture, forestry, coastal, and water resources will be adversely affected because of climate change. A devastating impact of climate change in India will be the rise in the sea level, resulting in the inundation of coastal areas. Coupled with these, the increase in cyclones accompanied by enormous volume of sea water would bring about mass devastation of human life as well as the economy. Estimate suggests that due to 1 m increase in the sea level, 7 million people would be displaced; about 5764 km^2 land and 4200 km stretch of roads would be lost (www.adb.org).

Indian economy has historically been an insignificant contributor to the global climate change. According to the INCCA (2010) report, India ranks fifth in aggregate GHG emissions in the world, behind USA, China, EU, and Russia in 2007. But the emissions of USA and China are almost four times that of India in the same year. Also, India's per capita carbon dioxide equivalent (CO2EQ) emission is 1.5 t/capita in 2007 which is roughly one fourth of the world average per capita emission of 4.5 t per annum (INCCA 2010). The main cause of CO_2 emission in India is the low-energy efficiency of coal-fired power plants, scarcity of capital, and the long lead time required to introduce advanced coal technologies. However, as India is now a fast-growing economy, its total emission is bound to grow rapidly. Restricting arise in or even lowering the carbon emission intensity can be a good strategy for India while it is on its way towards fast economic progress.

India signed the United Nations Framework Convention on Climate Change (UNFCCC) on June 10, 1992 and ratified it on November 1, 1993. It ratified the Kyoto Protocol on August 26, 2002 and hosted the Eighth Conference of the Parties (COP 8) in October 2002 in Delhi. There are number of projects underway directly aimed at reducing GHG emissions funded by the Global Environment Facility (GEF). Most of these projects are on renewable energy sources and biomass (Das et al. 2007). On the other hand, government of India has submitted its First National Communication (NATCOM I) to the UNFCCC in June 2004 and Second National Communication report (NATCOM II) in May 2010. This report, being brought out by INCCA, provides updated information on India's GHG emissions.

In India, policymakers are exploring various policy options that would limit carbon emissions. Stronger environmental measures encouraging use of clean fuel and improving energy efficiency are few of them (MoEF 2009). Not surprisingly, clean development mechanism (CDM) has generated much enthusiasm in India.

Parikh and Parikh (2002) point to a number of areas where India has reduced GHG emissions because of policies aimed at other goals. The gradual removal of energy subsidies and move towards free-market pricing for energy sources has been important in scaling back demand for coal in response to a 370% rise in the price of coal between 1980 and 1995. Electricity prices have risen even more over this period. Increased openness has meant that energy-efficient imported goods from white goods to motor vehicles have driven innovation in more energy-efficient Indian products.

The notable improvement in energy efficiency in India is well documented. It has been driven partly by policy and partly by price-induced incentives to conserve energy (Parikh and Parikh 2002). Also, the government of India has long promoted renewable energy sources. Other policies aimed at reducing local air pollution in the transport sector has also helped in lowering GHG emissions.

Again in the context of global climate change mitigation, the Annex-I countries, i.e., countries under Kyoto Protocol, committed to emission reduction targets, either voluntarily or by the protocol obligations, are allowed to trade among themselves the rights to emit GHGs (Bolin 1998). These rights are known as tradable permits. These rights materialize if at least some of the committed countries achieve additional GHG emissions abatement, over and above their committed abatement level. Countries for which the difference between assigned amounts and actual emissions is negative can potentially buy rights to emit from those for which the said difference is positive. It is obvious that tradable permits are market-based instruments meant to equalize marginal costs of abatement across countries.

However, there is considerable debate across the world about what can be the optimal policy response to mitigate GHG emissions. Economists, after weighing the costs and benefits, advocate a balanced mitigation program that starts with mild emission reduction targets which gradually increase in severity over the century. Scientists and environmentalists, in contrast, advocate more extreme near-term mitigation policies. What implications these two alternative approaches have for humankind is a burning research question. The balanced economic approach to the problem will address climate change with minimal reductions in economic growth, but is likely to impede human welfare in the long run. On the other hand, more aggressive near-term mitigation programs, recommended by the ecologists, would be harmful for economic growth, but helpful for the crumbling ecosystem on whose very existence depends the long-term survival and security of humanity (Stern 2007; IPCC 2010; Mendelshon 2009).

1.3 Past Studies

In the context of burgeoning GHG emissions and their extremely adverse impact on the economy, different national and international institutes as well as individual environmental economists are forging together to undertake scientifically rigorous impact analysis. One such study has been done by Darwin et al. (1995) which analyses how climate change might affect water supplies and the availability of cultivable land, which in turn would impact the total world production of goods and services. According to this study, world output of processed food would decline from 0.002 to 0.58%. In other words, the new temperature and precipitation patterns under climate change are likely to reduce the average productivity of the world's existing agricultural lands. Further, land-use changes that accompany climate-induced shifts in cropland, and permanent pasture are likely to raise additional social and environmental issues. Although water supplies are likely to increase for the world as a whole under climate change, shortages could occur in some regions. Finally, climate change is likely to affect the structure of agriculture and food processing in the USA most unfavorably.

In a similar vein, Pohit (1997) attempted to analyze the impact of climate change on India's agriculture using a 10 sector-10 region global CGE model, with India being one of its ten regions. The general conclusion that emanates from this study is that there could be substantial welfare implications for the Indian economy in general depending on how one accounts for the carbon fertilization effect.

Study by Kumar and Parikh (1996) focuses on assessment of the climate change impacts on Indian agriculture. The study is organized under two stages, namely, the physical impact assessment and the economic implications of such physical impacts. The future climate change scenarios have been developed using results from equilibrium experiments of General Circulation Models (GCMs), along with the observed climatic changes. In order to assess the physical impacts of climate change on agriculture, the study follows the crop simulation modeling approach and to translate these impacts into socio-economic impacts, the Agriculture, Growth, and Redistribution of Income Model (AGRIM) is used. Finally, the welfare implications are assessed in terms of equivalent incomes and population proportions in various expenditure classes of the economy. The results of the study indicate that wheat crop, grown generally in the winter season, is likely to be affected more than rice crop following climate change; CO_2 fertilization effects seem to reduce the effects of climate change dramatically. The study shows that substantial number of people move from higher income classes to lower income classes as a result of climate-change-induced shocks, and social welfare is adversely affected.

Mckinsey and Evenson (1998) estimate the impact of a rise in normal temperatures and of increases in rainfall levels for different regions. The study incorporates technology-climate interactions enabling an assessment of the climate friendliness of the Green Revolution in Indian agriculture. Technological gains during the period of Green Revolution are incorporated in the study by modeling three activities as endogenous variables, which are the development and diffusion of high yielding variety (HYV), expansion of multiple cropped areas, and the expansion of area under irrigation. The results of this study indicate that a 1 °C rise in temperature has negative impact on the HYV adoptions. This is most negative in Gujarat and is positive in some regions such as Andhra Pradesh, Orissa, etc. While the temperature impact on multiple cropping is positive on an average, that on irrigation is uniformly negative. Increased rainfall has negative but small effects on HYV adoption, irrigation intensity, and multiple cropping.

Ravindranath et al. (2006) assesses the impact of climate change on forests in India. This assessment is based on climate projections of regional climate model of the Hadley Center (HadRM3) using the A2 (740 ppm CO2) and B2 (575 ppm CO2) scenarios of special report on emissions scenarios and the BIOME4 vegetation response model. The main conclusion is that under the climate projection for the year 2085, 77% and 68% of the forested grids in India are likely to experience shifts in forest types under A2 and B2 scenarios, respectively. There are indications for a shift towards wetter forest types in the north-eastern region and drier forest types in the north-western region in the absence of human influence. Increasing atmospheric CO2 concentration and climate warming could also result in a doubling of net primary productivity under the A2 scenario and nearly 70% increase under the B2 scenario.

We turn now to the role of policy in climate change mitigation. To assess the impact of carbon taxes for the Indian economy, Gupta and Hall (1997) interweave a micro-level analysis of technological alternatives to reduce carbon emissions into a macro-econometric model. This interwoven model is especially helpful for assessing the effects of carbon tax financed investments in carbon-abating technologies. However, the issue of income distribution remains unaddressed as it is a macro model. Alternative policy scenarios built with the help of this model are compared mainly in terms of their implied carbon emissions and GDP, not with respect to their distributional implications.

Fischer-Vanden et al. (1997) employed a nine-sector computable general equilibrium (CGE) model of the Indian economy, based on the Indian module of the Second Generation Model (SGM) version 0.0. The SGM model is a typically neoclassical price driven CGE model, which is used to evaluate the effects of carbon taxes and participation in globally tradable permits regime on carbon emissions and economic growth. The SGM model assumes a single representative household and, hence, suppresses the income distribution aspect of an economy. However, it captures successfully the trade-off between carbon emissions and GDP growth. To evaluate the impact of carbon taxes on growth as well as distribution, the concerned model must include an endogenous income distribution mechanism. A model which does precisely that is that of Murthy, Panda, and Parikh (2007). This is an integrated top-down bottom-up model for the Indian economy which includes specific technological options and an income distribution module in an activity analysis framework. The model is multi-period and multi-sectoral and is formulated to be a programming model which allows for inter-temporal dynamic optimization. While the endogenous income distribution mechanism of the model is useful in computing the poverty ratios under different scenarios pertaining to carbon taxes and India's participation in an internationally tradable emission permits regime, the model, because it uses the activity analysis framework, is more like a planning model based on supply-side consistency rather than a market driven model which equates demand and supply through endogenous determination of prices. The CGE model of Ojha (2009) for India, however, embodies both an endogenous price system which balances demand and supply, and an endogenous income distribution module which enables the calculation of poverty ratios. It is therefore just appropriate for simulating the effects of carbon taxes and participation in a globally tradable emission permits regime on GDP and poverty.

Our review of literature brings forth research gaps. It is observed from the above review of literature that most of the studies in the subject of climate change impact analysis are for the vulnerable sectors of the economy like agriculture, forestry, fisheries, etc. On the other hand, there are some other sectors which may not be vulnerable but they may be making significant contributions to GHG emissions as well as on economic growth. For example, the manufacturing sector together contributes almost 27 % of total GHG emissions in the year 2006−2007 (INCCA 2010) and their share in GDP for that year is almost 18 % (CSO, National Accounts Statistics 2009). Therefore, growth in manufacturing sector will have significant impact on GHG emissions in India. In the year 2006−2007, thermal electricity sector alone contributes 37.8 % of total GHG emissions (INCCA 2010). Almost 86 % of energy in India comes from the thermal electricity sector (CSO, energy statistics 2009). So the growth of the Indian economy will depend on the growth of the thermal electricity sector, which in turn will lead to increase in GHG emissions in India. Again, as every sector is dependent upon and feeds every other sector through input-output linkages, the growth of a sector will have direct and indirect impacts on GHG emissions.

In India, policymakers are trying to provide different policy options for mitigating climate change impact in India. Energy efficiency improvement is one of them. Climate change experts are of the opinion that energy efficient technology will reduce future GHG emissions but not economic growth (MoEF 2009). On the other hand, the reduction in CO2EQ emission from economic activities by decoupling economic growth and GHG flow has been incorporated into the climate change mitigation policy agenda globally since the last two decades. To define appropriate policy interventions, a clear understanding of how emissions are generated and the economic and technological factors that influence the country's GHG profile is a must. This necessitates an in-depth study of this particular issue. Hence, this is another glaring research gap in this subject.

Apart from the energy efficiency improvement measures, there are three other policy measures for emissions abatement: command and control, carbon taxes, and participation in emissions trading (Ellis and Tirpak 2006). But the command-and-control measure has some serious limitations. Firstly, command-and-control measures have been shown to be statically and dynamically inefficient as compared to market-based instruments (MBIs) such as carbon taxes (Pearson 2000). Secondly, under a command-and-control measure, in order to reduce carbon emissions, it is the output of goods produced that has to contract as there is limited scope for substitution across (fuel) inputs. The loss in output in turn translates into a deadweight loss in welfare (Harrington and Morgenstern 2004). However, in case of MBIs like carbon taxes, the government can plough back the tax revenue productively to yield benefits for the economy over and above those resulting from lower emissions, thus, reducing the net loss in welfare. The government can also substitute the carbon tax for some other more distortionary tax and thus generate efficiency gains for the economy, i.e., reap double-dividend (Pearson 2000). Though the market-based incentive policies seemingly have direct and indirect advantages in India, their net impacts on the economy must be precisely evaluated and compared through a suitable empirical GHG emissions abatement model, so that they can be prioritized.

The role of clean development mechanism (CDM) and other mechanisms of the Kyoto Protocol in India's energy future are unclear (Murthy et al. 2007). The CDM allows late-industrializing countries to generate Kyoto permits that can be traded in an international market from the projects that otherwise would not have been undertaken and thereby reducing emissions below a baseline. A CDM project must be voluntary, generate "real, measurable, and long term benefits related to the mitigation of climate change," and generate "reduction in emissions that are additional to any that would occur in the absence of the certified project."[1] The main problem with the CDM is the problem of determining the baseline emissions that would otherwise have occurred as well as the amount of administrative cost involved in having CDM projects evaluated and approved. Probably the most attractive aspect of the CDM approach is the application to changes in land-use practice and forestation of degraded areas. However, India is already investing a lot of resources on reforestation independently of the CDM mechanism and it is difficult to define what is additional to baseline. Therefore, it seems that more direct policies aimed at changing the future composition of energy generation in India need to be considered.

---

[1] Page 12 of the test of the Kyoto protocol. Please see http://unfccc.int/resource/docs/convkp/kpeng.pdf.

Though researchers in India have made significant contributions in domestic and international climate policy formulation, there is a huge gap between policy prescription and implementation. Policy guidance in the form of a proper appraisal of the likely policy options with regard to their capability to benefit or harm the economy is urgently needed to make the ongoing policy debate conclusive and purposeful.

1.4   Study Goals

In the backdrop of the above research gaps in the area of climate change mitigation analysis, we have set the following objectives for our book:

1. To build an appropriate database and a methodology to assess the impact of economic growth on GHG emissions and that of sectoral output growth on energy demand, and employment.
2. To analyze empirically the factors driving changes in GHG emissions in India in order to bring out the importance of the emission intensities of different sectors in devising a climate change mitigation strategy of India.
3. To formulate an empirical model, such as a computable general equilibrium (CGE) model, to analyze the economy-wide impact of market-based policy instruments for mitigation in India.

1.5   Framework for Assessment

Assessments of impact of economic growth on climate change are not easy due to the complex relation between environment and economic activities (Jian 1996). The economy and environment interact with each other in myriad ways. To produce goods for consumption, a production process needs to depend on the environment to provide material resources and energy. The material resources and energy provided by the environment are transformed in the production and consumption processes to satisfy human wants, and the by-products discarded by the humans are then discharged back into the environment. Thus, environment is not only a provider of material and energy resources for the benefit of people inhabiting the earth, but also provides sink service for wastes generated from production and consumption. The environmental services are after all available in finite amounts and can therefore put a limit to the economic growth of the global economy.

There is eventually a limit to the environment's assimilative capacity by which it absorbs wastes discharged from the economy (Jian 1996). When the amount of wastes discharged into the environment is larger than the environment's assimilative capacity, environmental degradation occurs. The degradation of environmental quality has direct negative effects on both the utility of consumers and the stock of resources. The decrease in the quantity and quality of resources in turn has an indirect impact on consumer's utility or satisfaction by reducing productivity.

As most of the environmental problems can be directly attributed to the structure of production and consumption, it seems urgent and necessary and urgent to find ways to make these problems explicit within an accounting framework. In order to do this, it is important to develop a consistent accounting framework which will incorporate the economic as well as the environmental indicators. This work is an attempt to respond to the need to include, explicitly and directly, the two sets of indicators (economic and environmental indicators) into a unified and consistent framework which accounts for their relations to the economic system as a whole and provides the basis for diagnoses and eventually for policy-making.

However, the conventional national accounting system is not accommodative towards environmental indicators. Including environmental indicators in a comprehensive national accounting system, thus, poses an enormous challenge from a methodological standpoint. The ensuing analysis is best seen as a modest beginning in meeting this challenge. In order to overcome some of the limitations of the conventional national accounting system, more comprehensive systems have been developed, among which the traditional and extended input output (IO) tables and their generalized form as Social Accounting Matrices (SAMs) are the most prominent.

A SAM depicts the entire circular flow of income for an economy in a (square) matrix format. It shows production leading to the generation of incomes which in turn are allocated to institutions—households, enterprises, government, rest of the world, etc. The disposable incomes (which is nothing but earned incomes net of direct taxes) of these institutions are either spent on products or saved. Expenditures by institutions create demands which are met by domestic production from domestic industries as well as from imports. The advantage of incorporating both the economic and environmental indicators in a common social accounting matrix framework is that their interrelations can become more transparent for policymakers. Therefore, the extension of a conventional SAM to include environmental indicators can be considered as the first logical step in the efforts to simultaneously account for the interrelationships between economic and environmental activity.

Another advantage of SAM is its enabling of multiplier analysis. With the help of the SAM multiplier, we can analyze the direct, indirect, and induced impact of exogenous factors in the economy. Therefore, if we interrelate the economic indicators with the environmental indicator in a SAM framework, then that environmental social accounting matrix (ESAM) will help us to analyze the impact of economic growth on climate change.

To analyze the factors which are responsible for the GHG emissions, many researchers follow the structural decomposition analysis (SDA) to distinguish the factors and their impacts on GHG emissions (Mukhopadhyay 2001). As suggested in the literature, we need an integrated database which will provide data on IO coefficients as well as data on GHG emissions. The ESAM will eminently serve that purpose.

Finally, we apply the method of CGE model to analyze the impact of carbon mitigation policies in India. This CGE model used here is an integrated top-down model which assimilates the economy and environment in a single modeling framework. Unlike IO type models, CGE models are characterized by nonlinear and price-endogenous features and the inclusion of resource constraints. They thus effectively reflect real-world problems. Devarajan (1988) specifically lists three reasons why CGE models, rather than other types of economy-wide models, are preferred for policy analysis. The first reason is that price matters. CGE models are distinguished by their price endogenous features. Prices and quantities are determined simultaneously in simulating the results of an external shock or a policy change. The second reason is that interactions matter. CGE models are specifically designed to include many markets (such as goods and factor markets), many institutions (such as firms, households, and government), and their interactions. The third reason is that economic structure matters. CGE model focuses on the issue of economic structure. In addition to its internalization of market mechanisms, the CGE approach leaves room for nonmarket activities. Therefore, due to these advantages, the CGE approach is more capable of simulating the results of a policy change or an external shock than are other previous predecessor models such as IO models.

1.6   Scheme and Scope of This Study

Following this introductory chapter, the rest of this book is organized as follows: Chapter 2 provides the concepts and construction of the social accounting matrix for India. Chapter 3 describes the method of constructing ESAM with the environmental indicators which are significant for climate change analysis. Chapter 4 describes the estimation of SAM multiplier and its application to show the impact of economic growth on climate change. Chapter 5 illustrates IO structural decomposition analysis to analyze the factors determining change in GHG emissions in India. In Chap. 6, we have described the structure of the India CGE model for climate change mitigation policy analysis and its underlying assumptions. The results of this CGE model are described in Chapter 7, and, thereafter, Chapter 8 summarizes the study and provides concluding remarks with hints about future scope of research.

References

Agarwal A, Narain S (1991) Global warming in an unequal world: a case of environmental colonialism. Center for Science and Environment, New Delhi
Bach SM, Kohlhaas B, Meyer BP, Welsch H (2001) The effects of environmental fiscal reform in Germany: a simulation study. Energy Policy 30(9):803–811
Blackman A, Harrington W (2000) The use of economic incentives in developing countries: lessons from international experience with industrial air pollution. J Environ Develop 9(1):5–44
Bolin B (1998) The Kyoto negotiations on climate change: a science perspective. Science 279:330–331
CSO (Central Statistical Organization) (2009) Energy statistics, ministry of statistics and program implementation, Government of India

Darwin R, Tsigas M, Lewandrowski J, Ranese A (1995) World agriculture and climate change—economic adaptations. Agricultural economic report no 703, Natural Resources and Environment Division, Economic Research Services, United States Department of Agriculture, Washington, DC, USA

Das S, Mukhopadhyay D, Pohit S (2007) Role of economic instruments in mitigating carbon emissions: an Indian perspective. Econ Polit Wkly XLII(24):2284–2291

Devarajan S (1988) Natural resources and taxation in computable general equilibrium models of developing countries. J Policy Model 10(4):505–528

Ellis J, Tirpak D (2006) Linking GHG emissions trading systems and markets. OECD, COM/ENV/EPOC/IEA/SLT (2006) 6

Fischer-Vanden KA, Shukla PR, Edmonds JA, Kim SH, Pitcher HM (1997) Carbon taxes and India. Energy Econ 19:289–325

GEA (2012) Global energy assessment: toward a sustainable future. International Institute for Applied Systems Analysis, Vienna, Austria and Cambridge University Press, Cambridge, UK and New York, NY, USA

Gupta S, Hall S (1997) Stabilizing energy related CO_2 emissions for India. Energy Econ 19(1):125–150

Harrington W, Morgenstern DR (2004) Economic incentives versus command and control—what is the best approach for solving environmental problems? Resources for the Future, Fall/Winter, 2004

INCCA (Indian Network on Climate Change Assesment) (2010) India: greenhouse gas emission 2007. Ministry of Environment and Forests, Government of India

IPCC (Intergovernmental Panel for Climate Change) (2010) The fourth assessment report. Cambridge University Press, Cambridge

Jian X (1996) Environmental policy analysis, a general equilibrium approach. Avebury, England

Kumar KS, Parikh J (1996) Potential impacts of global climate change on Indian agriculture, Communicated to Global Environment Change, 1996

Mckinsey JW, Evenson RE (1998) Technology-climate interactions: was the green revolution in India climate friendly?" In: Dinar AR, Mendelsolm RE, Parikh J, Sanghi A, Kumar K, McKinsey J, Lonnergan S (ed) Measuring the impact of climate change on Indian agriculture, World Bank technical Paper no 402, Chap. 6. The World Bank, Washington, DC

Mendelsohn R (2006) Climate change impacts on agriculture. In: Evenson R, Pingali P, Schultz, P (eds.) Handbook of agricultural economics: agricultural development, vol 3. Chapter 19

Mendelshon R (2009) Climate change and economic growth. Working paper no 60, Commission of growth and development, World Bank

MoEF (2009) India's greenhouse gas emission inventory—a report of five modeling studies. Ministry of Environment and Forests, Government of India

Mukhopadhyay K (2001) An empirical analysis of sources of CO_2 emission changes in India. Asian J Energy Environ 2(3–4):233–271

Murthy, Panda, Parikh (2007) CO_2 emission reduction strategies and economic development of India. Margin 1(1):85

Ojha VP (2009) Carbon emission reduction strategies and poverty alleviation in India. Environ Develop Econ 14(3):323–348

Parikh J, Parikh K (2002) Climate change: India's perceptions, positions, policies and possibilities, OECD, Paris

Parikh J, Parikh K, Gokran S, Painuly JP, Saha B, Shukla V (1991) Consumption patterns: the driving force of environmental stress. Paper presented at the United Nations Conference on Environment and Development (UNCED), IGIDR, Monograph

Pearce DW, Cline WR, Achanta AN, Fankhauser S, Pachuri RK, Tol RSJ, Vellinga (1996) The social costs of climate change: greenhouse damage and the benefits of control In: Bruce JP, Lee H, Haites EF (eds) Climate change 1995: economic and social dimensions—contribution of working group III to the second assessment report of the intergovernmental panel on climate change. Cambridge University Press, Cambridge

Pearson CS (2000) Economics of the global environment. Cambridge University Press, Cambridge

Pohit S (1997) The impact of climate change on India's agriculture: some preliminary observations. Proceeding of the 20th International Conference of the International Association for Energy Economics, 22–24th January, Delhi, India

Ravindranath NH, Joshi NV, Sukumar R, Saxena A (2006) Impact of climate change on forests in India. Curr Sci 90(3):354–361

Sathaye J, Shukla PR, Ravindranath NH (2006) Climate change sustainable development and India: global and national concerns. Curr Sci 90(3):314–325

Sharma S, Bhattacharya S, Garg A (2006) Greenhouse gas emission from India: a perspective. Curr Sci 90(3):326–333

Stern N (2007) The economics of climate change—a stern review. Cambridge University Press, Cambridge

Chapter 2
Social Accounting Matrix of India: Concepts and Construction

Social accounting matrix (SAM) is a technique related to national income accounting, providing a conceptual basis for examining both growth and distributional issues within a single analytical framework in an economy. It can be seen as a means of presenting in a single matrix the interaction between production, income, consumption, and capital accumulation. In this Chapter, we describe the concept and methodology for construction of a SAM for India. The novelty of this SAM is a detailed account of disaggregated energy sectors, electricity sectors, energy intensive sectors, and biomass as an alternative source of fuel. Further, this is one of the latest SAMs for India with such a high level of disaggregation.

2.1 Concept and Structure of SAM

A social accounting matrix is simply defined as a single entry accounting system whereby each macroeconomic account is represented by a column for outgoings (payments) and a row for (receipts) incomings (Round 1981). It is represented in the form of a square matrix with rows and columns, which brings together data on production and income generation as generated by different institutional groups and classes on the one hand and data about expenditure of these incomes by them on the other. In a SAM, incomings are indicated as receipts for the row accounts in which they are located and outgoings are indicated as expenditure for their column accounts. Since all incomings must be, in a SAM, accounted for by total outgoings, the total of rows and columns must be equal for a given account.

SAM is essentially a database, including both social and economic data for an economy for an accounting year. The data sources for a SAM come from input-output (IO) tables, national income statistics, and household income and expenditure statistics. Therefore, a SAM is broader database than an IO table and typical national account, showing more detail about all kinds of transactions within an economy. An IO table generally records economic transactions alone irrespective of the social background of the transacting actors. A SAM, on the contrary the national accounts, "…attempts to classify various institutions to their socio-economic backgrounds instead of their economic or functional activities" (Chowdhury and Kirkpatrick 1994, p. 58).

B. D. Pal et al., *GHG Emissions and Economic Growth*,
India Studies in Business and Economics, DOI 10.1007/978-81-322-1943-9_2,
© Springer India 2015

At this point, it would be good to describe the various components of a hypo-thetical SAM in a schematic diagram (Table 2.1). As this table shows, there are four agents in the hypothetical economy, namely, the households, the private corpo-rate, the pubic nondepartmental enterprises and the government. Here, interindustry flows are presented by A11. This table indicates that the factor incomes generated through production process (A21) are transferred to institutions according to the ownership of their factors of production (A32, A42, A52, and A62). In addition, a household gets its income from current transfers from the government as well as from interest on public debt (A36) and the net current transfers from the rest of the world (ROW; A39). The households spend on consumption of goods and services (A13) and pay income taxes (A63) and indirect taxes on purchase (A73) and they keep the residual income as savings (A83).

The income of the private corporate sector comes from its operating profit (A42) and interest on holding public debt (A46). After payment of corporate tax (A64), the residual is savings (A84).

The receipts and expenditures of the other two institutions, public-nondepartmen-tal enterprises and government administration including departmental enterprises, are specified in this table. The income of the first category is only the operating surplus (A52) which is also its saving (A85). The receipts of the government con-sists of income from its enterprises (A62), direct taxes paid by the households and private corporations (A63, A64), the total indirect taxes generated within the econ-omy (A67) and the net capital transfer from ROW (A69). On the other hand, its outlay includes its final consumption expenditure on goods and services (A16), transfers and interest payments to households (A36) and interest payments to pri-vate corporate sector (A46).

The receipts of the capital account are from the net savings of the different insti-tutions (A83, A84, A85, and A86), foreign savings (A89), and depreciation (A82). The expenditure is equal to gross domestic capital formation (A18) and indirect taxes paid on purchases of the investment goods (A78).

The ROW represents the equality between foreign exchange expenditures on the one hand and foreign exchange earnings on the other. Foreign exchange expenditure equals imports (A91). On the other hand, foreign exchange earnings equals sum of exports (A19), net factor income from abroad (A29), net current transfers (A39), net capital transfer (A69), net export taxes (A79), and foreign savings (A89).

2.2 Purpose of Constructing SAM

India has been an early leader in SAM-based model users. To the best of our knowl-edge, Sarkar and Subbarao (1981) constructed the first SAM for India back in the 1980s, which provides the consistent database for their computable general equilib-rium (CGE) model. Subsequently, a number of SAMs are constructed over the years by the different researchers. In the following Table 2.2, we have described a brief outline of these various SAMs and their salient features.

Table 2.1 A schematic social accounting matrix (SAM) for India. (Source: Pradhan et al. 2006)

Expenditures / Receipts	Production account	Factors of production	Households	Private corporate	Public nondepartmental	Government	Indirect taxes	Capital account	Rest of the world (ROW)
Production account	Input-output table A11		Private consumption A13			Government consumption A16		Investment A18	Exports A19
Factors of production	Value added (VA) A21								Net factor income A29
Households		VA income A32				Government transfers, interest on debt A36			Net current transfers A39
Private corporate		Operating profits A42				Interest on debt A46			
Public nondepartmental		Operating surplus A52							
Government		Income from enterprises A62	Income and wealth taxes A63	Corporate taxes A64			Total indirect taxes A67		Net capital transfer A69
Indirect taxes	Taxes on intermediate A71		Taxes on purchases A73			Taxes on purchases A76		Taxes on investment goods A78	Taxes on exports A79
Capital account		Depreciation A82	Households savings A83	Corporate savings A84	Public nondepartmental savings A85	Government savings A86			Foreign savings A89
Rest of the world (ROW)	Imports A91								

Table 2.2 Stylized facts of social accounting matrices (SAMs) of India. (Source: Authors' collection)

Serial no.	Name of researchers and their SAM-based study	Salient features of SAM
1.	Sarkar and Subbarao (1981)	*Base year*: 1979–1980 *Sectors (3 in all)*: agriculture, industry, and services *Agents*: nonagricultural wage income class, nonagricultural nonwage income class, agricultural income class, and government
2.	Sarkar and Panda (1986)	*Base year*: 1983–1984 *Sectors (6 in all)*: agriculture (2), industry (2), infrastructure, and services *Agents*: nonagricultural wage income class, nonagricultural nonwage income class, agricultural income class, and government
3.	Bhide and Pohit (1993)	*Base year*: 1985–1986 *Sectors (6 in all)*: agriculture (2), livestock and forestry, industry (2), infrastructure, and services *Agents*: government, nonagricultural wage income earners, nonagricultural profit income earners, and agricultural income earners
4.	Pradhan and Sahoo (1996)	*Base year*: 1989–1990 *Sectors (8 in all)*: agriculture (2), mining and quarrying, industry (2), construction, electricity combined with water and gas distribution, and services (3) *Agents*: government, agricultural self-employed, agricultural labor, and nonagricultural self-employed and other labor
5.	Pradhan et al. (1999)	*Base year*: 1994–1995 *Sectors (60 in all)*: agriculture (4), livestock products (2), forestry sector, mining (4), manufacturing (27), machinery and equipment (6), construction, electricity, transport (2), gas and water supply, other services (11) *Agents*: government, self-employed in agriculture (rural and urban), self-employment in nonagriculture (rural and urban), agricultural wage earners (rural and urban), other households (rural and urban), private corporate, and public nondepartmental enterprises
6.	Pradhan et al. (2006)	*Base year*: 1997–1998 *Sectors (57 in all)*: agriculture (4), livestock products (2), forestry, mining, manufacturing (27), machinery and equipment (6), construction, electricity, transport (2), gas and water supply, other services (11) *Agents*: government, self-employed in agriculture (rural and urban), self-employment in nonagriculture (rural and urban), agricultural wage earners (rural and urban), other households (rural and urban), private corporate, and public nondepartmental enterprises
7.	Sinha et al. (2007)	*Base year*: 1999–2000 *Sectors (13 in all)*: agriculture (informal), formal manufacturing (9), construction (informal), other services (formal and informal), and government service *Agents*: casual labor (rural and urban), regular wage earner (rural and urban), own account worker (rural and urban), employer (rural and urban), and government

Table 2.2 (continued)

Serial no.	Name of researchers and their SAM-based study	Salient features of SAM
8.	Saluja and Yadav (2006)	*Base year*: 2003–2004 *Sectors (73 in all)*: agriculture (12), livestock products (4), forestry, mining (4), manufacturing (28), machinery and equipment (7), construction, energy, gas distribution, water supply, transport (2), other services (10) *Agents*: five rural households' expenditure classes, five urban households' expenditure classes, private corporation, public enterprises, and government
9.	Pal et al. (2012)	*Base Year*: 2003–2004 *Sectors (85 in all)*: agriculture (19), livestock products (1), forestry, mining (9), manufacturing (32), construction, electricity (3), biomass, water supply, transport (5), other services (12) *Agents*: five rural households' occupation classes, five urban households' occupation classes, private corporation, public enterprises, and government

As Table 2.2 shows, the SAM constructed by Pal et al. for the year 2003–2004 provides detailed description about primary energy, biomass, electricity, and transport sector. Prior to this SAM the available SAMs, especially before 1996, were highly aggregated in nature and most of them have three household classes. These SAMs are old and their socioeconomic classifications are based on data for the year 1970. To some extent, SAM constructed in the post 1996 period addresses these shortcomings. However, the article published by Pal et al. does not provide detail description about the methodology for constructing such detail SAM for India. Moreover, the base year of this SAM is 2003–2004, whereas the government of India had published the IO table for the year 2006–2007 also (Central Statistical Organization, CSO 2010).

Since our interest of SAM is for climate change policy analysis, we have planned to present a detail methodology of constructing a SAM for India with detail description of primary energy, biomass, electricity, and transport sectors. Moreover, we have decided to construct a SAM for the more recent year 2006–2007 than the 2003–2004 SAM as available in Pal et al (2012). However, looking at the data availability, our need, and time constraint, we have decided to construct a SAM of 35 sectors of the economy, 3 factors of production, and 9 categories of occupational households. The description of the sectors of our SAM and its concordance map with 130 sectors of IO flow table is shown in Table 2.3.

As Table 2.3 shows, some of the sectors of this SAM match with the sectors of the 130-sector IO table of the year 2006–2007. But the important aspect of our SAM is the decomposition of electricity sector into three separate sectors viz. hydro, nuclear, and non-hydro. The non-hydro energy sector includes thermal, wind power, solar energy, etc. However, given India's energy balance, thermal is the main constituent of this group (CSO, Energy Statistics 2007). Another salient feature of this SAM is the incorporation of biomass. The biomass is an alternative source of commercial and domestic fuel.

Table 2.3 Mapping between social accounting matrix (SAM) sectors and sectors of input-output (IO) table. (Source: Authors' estimate)

Serial no.	Sector code	Sectors for SAM	Sectors of IO table
1	PAD	Paddy rice	1
2	WHT	Wheat	2
3	CER	Cereal, grains, etc., other crops	Part of (3–7,18,19, 20)
4	CAS	Cash crops	8,9,10–17
5	ANH	Animal husbandry and production	Part of (21, 22, 23, 24)
6	FOR	Forestry	Part of 25
7	FSH	Fishing	26
8	COL	Coal	27
9	OIL	Oil	29
10	GAS	Gas	28
11	MIN	Minerals not elsewhere classified	30–37
12	FBV	Food and beverage	Part of (38–45)
13	TEX	Textile and leather	46–54, 59, 60
14	WOD	Wood	56
15	PET	Petroleum and coal production	63,64
16	CHM	Chemical, rubber, and plastic production	58,61,62,65,66, 69–73
17	PAP	Paper and paper production	Part of 57
18	FER	Fertilizers and pesticides	67,68
19	CEM	Cement	75
20	IRS	Iron and steel	77,78, 79
21	ALU	Aluminum	80
22	OMN	Other manufacturing	55, 74, 76, 81, 82, 95–105
23	MCH	Machinery	83–94
24	HYD	Hydro	107
25	NHY	Thermal	107
26	NUC	Nuclear	107
27	BIO	Biomass	Part of (3–7,18,19, 20), part of (21, 22, 23, 24), part of 25, part of (38–45), part of 57
28	WAT	Water	108
29	CON	Construction	106
30	LTR	Land transport	110, 113
31	RLY	Rail transport	109, 113
32	AIR	Air transport	112, 113
33	SEA	Sea transport	111, 113
34	HLM	Health and medical	122
35	SER	All other services	114–121, 123, 124–126, 127–130

The description of 130 sectors of IO flow table is given in Appendix 1

We have considered four economic agents in our proposed SAM viz. households, government, public nondepartmental enterprises, and private corporate. Moreover, the households are decomposed into nine occupational households group. This household's classification is based on the household's classification given by National Sample Survey Organization (NSSO 2008). Thus, our SAM consists of 12 economic agents. The description of these 12 economic agents is given in Table 2.4.

Table 2.4 Description of economic agents. (Source: NSSO 2008)

Agent code	Description
RNASE	Rural nonagricultural self-employed
RAL	Rural agricultural labor
ROL	Rural other labor
RASE	Rural agricultural self-employed
ROH	Rural other households
USE	Urban self-employed
USC	Urban salaried class
UCL	Urban casual labor
UOH	Urban other households
PVT	Private corporate
PUB	Public nondepartmental enterprises
GOV	Government

Once we know the sectoral description of our proposed SAM of India, our next task is to construct this SAM for the Indian economy. In the following sections, we describe in detail the methodology of constructing the 35-sector SAM of India for the year 2006–2007.

2.3   Methodology of Construction of SAM

The core of a SAM is an IO table. CSO has prepared 130-sector commodity X industry absorption matrix and 130-sector industry X commodity make matrix and sectors commodity X commodity IO flow matrix for the year 2006–2007.[1] Since we have made an attempt to construct a SAM in a commodity X commodity framework for the year 2006–2007, we have considered commodity X commodity IO table for India for the year 2006–2007.

However, the concordance map of our sectors with the 130 sectors of IO table as shown in Table 2.3 indicates that sectors like biomass, hydroelectricity, nuclear electricity, and non-hydroelectricity do not have a one-to-one mapping with 130 sectors. So to complete our exercise, we need to construct rows and columns for these sectors, and the procedure is given below.

2.3.1   Expansion of Electricity Sector (Hydro, Non-hydro, and Nuclear)

The Hydro Power Corporation Limited of India (NHPCL) and Nuclear Power Corporation of India Limited (NPCIL) publish their annual accounts every year. The information from these reports has been used to construct the columns of hydro- and

---

[1] www.mospi.nic.in extracted on August 2010.

nuclear-electricity sectors. Once we get the column of hydro- and nuclear-electricity sectors, we subtract them from aggregate electricity sector to obtain the column of non-hydroelectricity sector. Thus, we get the total output (i.e., column sum) of each electricity sector and their corresponding share with aggregate electricity sector's output. We apply this share to the row of aggregate electricity sector to obtain the rows of these electricity sectors separately.

2.3.2   Construction of Biomass Sector

The biomass supplies originate in (1) agricultural residuals, (2) animal husbandry residual, (3) firewood, (4) food and beverages industry residuals, and (5) paper and paper industry residuals.[2] According to the national accounting methods, only the 5 % of agricultural residuals are taken as biomass and rests of the part are considered to be consumed by the entire livestock population (CSO 1989).

The use of biomass can be divided into two parts viz. noncommercial purpose and commercial purpose (Fritz and Steininger 1998). Noncommercial biomass is mainly used as cooking fuels, which comes from agriculture, animal husbandry, and forestry sectors. On the other hand, commercial biomass use has two parts viz., commercial nonmodern biomass use and modern biomass use. The commercial nonmodern biomass use considers the use of charcoal and commercial fuel wood and these are available from forestry sector. The paper and sugar industry are the main contributors of modern biomass. This modern biomass is used mainly for bio fuels. Apart from this, the biomass can also be used for the production of chemicals, plastic, as well as reducing agent for steel production (charcoal) and for construction purpose.

Now the data on agricultural sector residuals, animal husbandry residuals, and firewood are available from National Accounts Statistics (NAS) of India. On the other hand, Annual Survey of Industries (ASI) gives data on commercial nonmodern biomass use as well as modern biomass use statistics for the year 2000–2001. We use ASI data on biomass use to obtain total biomass output originating from industries for the year 2000–2001. Again, this industrial biomass is originated from paper industry and food and beverages industry. We use their ratio of output of the year 2000–2001 to split up this industrial biomass according to their origin. To obtain the industrial biomass output for the year 2006–2007, we first estimate the share of paper and food and beverages industries in biomass production and apply this to their total output of the year 2006–2007. Thus, we get the total biomass output of the year 2006–2007.

Once we have derived total output data on biomass, we need to estimate its row and column. But, as data is not available on input structure of biomass sector, it is very difficult to make a column of this sector. In this case, we have considered these five types of residuals viz., agricultural residual, animal husbandry residual,

---

[2] See http://edugreen.teri.res.in/explore/renew/biomass.htm.

firewood, food and beverages industry residuals, and paper industry residuals as by-products of their mother sectors and apply the input structure of their mother sector to these by-products (Pradhan et al. 2006).

To make a row of biomass sector, we first make separate rows for the agricultural residuals, animal husbandry residuals, firewood, and industrial residuals. As we mentioned earlier, the agricultural and animal husbandry residuals are mainly used as cooking fuel. But in the agricultural production process, these residuals are also used as organic fertilizer. So, we distribute these residuals into the agriculture-related sectors on the basis of the rows of their mother sectors and the rest of the part is treated as Private Final Consumption Expenditure (PFCE). Since the firewood is mainly used as cooking fuel, we have treated the firewood output as PFCE. As the industrial biomass use statistics for the year 2000–2001 is available from the ASI, we use this ratio to distribute the industrial biomass output of the year 2006–2007. The point to be noted is that these residuals are included under their corresponding mother sectors described in IO table 2006–2007. So, after obtaining the columns and rows of these residuals, we have subtracted them from the columns and rows of their corresponding mother sectors to obtain the independent rows and columns of these residuals. At last, we add up these columns and rows to get a column and row of biomass sector. In this way, we have extended the IO table of the year 2006–2007 with the three electricity sectors and one biomass sector.

On the other hand, the IO table of the year 2006–2007 gives accounts for four transport activities and one separate account for services incidental to transport sectors. But in our SAM, there is no separate account for the services incidental to transport activities; rather it has been included in all these transport activities. Therefore, to complete the SAM of the year 2006–2007, we have to include the row and column of these services incidental on transport sector into four transport sectors and the method is given below.

2.3.3 Disaggregation of "Services Incidental to Transport" Sectors

The services incidental on transports described in IO table 2006–2007 comprises packing, crating, operations of travel agencies, etc. These services are associated with shipping, air, railways, and road transport. In our SAM, there is no separate account of this sector; rather it has been merged with sea, air, railways, and road transport sectors. Therefore, we are to break-up the row and column of this services incidental on transport sector according to transport activities in which these services are associated. To do this, we have applied the share of output of each transport sector to the row and column of this services incidental on transport sector. The rows and columns thus obtained are added to the row and column of sea, air, railways, and road transport sectors, and hence, we get the four transport sectors exactly matching with transport sectors described in our SAM.

2.3.4 Aggregation of IO Table

After expanding the 130-sector IO table 2006–2007, we obtain a 132-sector IO table of India for the same year. To be specific, the added three new sectors are hydroelectricity, nuclear electricity, and biomass, and the sector describing services incidental to transport have been merged with road, rail, sea, and air transport. According to our sectoral scheme, we aggregate the 132-sector IO table to obtain 35-sector IO table. The next step is to extend the accounts of 35-sector IO table into accounts for the 35-sector SAM. This involves primarily decomposition of gross value added into depreciation, wage and nonwage income, and of PFCE and personal income into economic categories of households. The next section describes the methodology of same.

2.3.5 Extension of IO Table to SAM

We describe below the methodology along with the data sources for the decomposition of gross value added into depreciation, wage, and nonwage income. The other relevant accounts for the SAM are also discussed in this section.

Decomposition of Gross Value Added into Depreciation, Wage, and Nonwage Income The decomposition of gross value added into wage (including imputed) and nonwage income (capital and land income) has been done for 35 sectors of the economy for 2006–2007. The sources of data and methods used are given below according to the broad sectors of our SAM.

Agriculture and Allied Activities (Sectors 1–7 of SAM) The aggregate net value added (NVA) for agriculture, forestry, and fishing sector is available from NAS. We first calculate the NVA separately for these sectors by using the depreciation to gross value added (GVA) ratio for the entire agricultural sector, as available from the NAS. As we have considered four sectors under agriculture and an animal husbandry sector as separate activity in our SAM, we divide the aggregate NVA of agricultural sector into these five sectors on the basis of above-mentioned method. The NAS also gives the breakdown of the NVA into compensation to employees (CE) and operating surplus/mixed income (OS). The CSO divides the mixed income under the unorganized part of agriculture and animal husbandry sectors into the income of family labor and operating surplus, for which data are available for the period 1980–1981 to 1999–2000 (CSO 2008). We use the proportion of 1999–2000 to disaggregate the mixed income of 2006–2007 into the above two categories. The wage income due to family labor, obtained in this way, has been added to the actual wage income from the organized and unorganized components to get the total income due to labor. Having obtained the total labor income of the aggregate agriculture sector, we distribute this labor income into four agriculture-related sectors by using the proportion of NVA as obtained earlier. The remaining part of the net domestic product is the operating surplus. We add depreciation with this operating surplus to obtain the sector specific capital income.

Since the relevant data for forestry and fishing is not available, the mixed income of unorganized part of this sector is divided into wage income and operating surplus by using the same ratio as in agriculture. Finally, the total value added in each of these sectors is divided into its components by applying the same method as used for agriculture.

Again, as the land is an important part of the capital in crop husbandry, nonwage income of this sector is divided into income from land and other capital. This is done by using information from the databases available with the Ministry of Agriculture, Government of India (http://eands.dacnet.nic.in/).

Mining and Quarrying (Sectors 8–11 of SAM) In our SAM 2006–2007, there are four sectors under mining. They are namely coal, crude petroleum, natural gas, and minerals not elsewhere classified. The IO table 2006–2007 gives the GVAs of these four sectors. We need to decompose these GVAs into depreciation, wage, and nonwage income. NAS provides data of GVA separately for NVA and depreciation of the aggregate mining and quarrying sector and also gives the decomposition of NVA into wage and nonwage income of the same sector. We first apply the ratio between GVA and depreciation of the aggregate mining and quarrying sector to the independent GVA of coal, crude petroleum, and minerals sectors to obtain the NVAs of these sectors. Next, we decompose these NVAs into wage and nonwage income by using the ratio of wage and nonwage income of the aggregate mining and quarrying sector. The depreciation of each of these sectors is added with their nonwage income to obtain the gross capital income of these sectors, respectively.

Next, we have to estimate the wage and nonwage income of the natural gas sector. As the natural gas sector of the IO table 2006–2007 is merged with gas distribution sector, we have subtracted the GVA of gas distribution sector from the GVA of natural gas sector of the IO flow table 2006–2007. Once we have obtained the GVA of independent natural gas sector, we apply the same method as applied for the coal, crude petroleum, and minerals sectors to obtain the depreciation, wage, and nonwage income of the independent natural gas sector. Next, we need to decompose the GVA of gas distribution sector into depreciation, wage, and nonwage income. The data on depreciation and NVA of the gas distribution sector is directly available from the NAS. The NAS also divides the NVA into wage and nonwage income of the organized as well as unorganized part of the combined electricity, gas and water supply sector. We have subtracted the wage and nonwage income of the electricity and water supply sectors from the same of the organized part of the combined sector to obtain the wage and nonwage income of the organized gas distribution sector. The wage and nonwage income of unorganized part of the combined sector is treated as wage and nonwage income of the gas distribution sector. The unorganized part of the combined electricity, gas, and water supply sector mainly deals with gas distribution sector. The wage and nonwage income thus obtained for the gas distribution sector is added with the same of independent natural gas sector. Finally, we add up the depreciation of independent natural gas and gas distribution sectors with the nonwage income of the natural gas sector of IO table 2006–2007 to obtain the gross capital income of the gas sector.

Manufacturing Industries (Sectors 12–23 of SAM) The output of manufacturing industries comprises of the outputs of the registered and unregistered sectors. The GVA at 2-digit level of National Industrial Classification (NIC) for 2006–2007 given in NAS is divided into wage and nonwage income on the basis of ASI data as available for 2003–2004. For unregistered manufacturing, the NSSO gives estimate of GVA, emoluments, number of hired, and total workers for the year 2005–2006 (NSSO 2008). In case of self-employed workers, the imputed values based on the data for hired workers are used. Using the proportion of different components of GVA for 2000–2001 for the registered sectors to the 2006–2007 GVA of the unregistered sector, we obtain the components of GVA at the 2-digit level industrial classification for the year 2006–2007. Adding these values for registered and unregistered sector, we obtain the components of the GVA for the entire manufacturing sector at 2-digit level. Lastly, we use the ratios for each 2-digit level industrial group for all the sectors under that group to obtain the wage and nonwage incomes for different sectors under manufacturing. Since output of sectors, like cement, fertilizers, mainly come from the organized sector, we have directly used the ASI ratios.

Transport (Sectors 30–33 of SAM) The estimate of wage and nonwage income for railway transport is obtained directly from NAS. But the data on wage and nonwage income for organized part of the other transport sector are available from the NAS. Since the organized transport sector contains combined of land transport, sea transport, and air transport, the NVA of these sectors are divided by the same ratio that is used for entire organized transport sector. To divide the GVA into wage and nonwage income of the unorganized transport sector, we use the information available from Enterprise Survey of CSO 1993–1994 and add this income to land transport sector.

Other Services Sector (Sectors 34 and 35 of SAM) Besides these transport sectors, there are another two services sector described in our SAM, i.e., health and medical services and other services. The NVA data on different service sectors are available from NAS. We add up the wage and nonwage income of different services to obtain the value added of these two sectors. We have treated mixed income as nonwage income of service sectors.

Electricity (Sectors 24–26) The electricity sector comprises three different electricity sectors (i.e., hydro, non-hydro, and nuclear) in our SAM. The data on GVA and depreciation for the hydro- and nuclear-electricity sectors are available, respectively from the annual accounts of NHPCL and NPCI. On the other hand, NVA and depreciation of aggregate electricity sector is available from NAS. Since we have estimates of NVA and depreciation of hydro- and nuclear-electricity sectors, we can derive NVA and depreciation of the non-hydroelectricity sector.[3] To decompose the NVAs of these electricity sectors into wage and nonwage income, the following procedure is adopted. NAS gives the decomposition of NVA for the aggregate electricity, gas, and water supply sectors. Using this ratio of wage and

---

[3] Note: NVA (non-hydro)=NVA (electricity)—NVA (hydro)—NVA (non-hydro), depreciation (non-hydro)=depreciation (electricity)—depreciation (hydro)—depreciation (non-hydro).

nonwage income, we obtain the wage and nonwage income of these three electricity sectors. Next, we add the depreciation with the nonwage income of these electricity sectors to obtain the total nonwage income of these separate electricity sectors.

Water Supply (Sector 28 of SAM) The decomposition of GVA into NVA and depreciation is given in NAS for water supply sector. However, NAS does not provide the data on wage and nonwage income of the water supply sector separately. To decompose the NVA of water supply sector, we have used the ratio of wage and nonwage income of the combined electricity, gas, and water supply sector. The nonwage income of the water supply sector thus obtained is added with the depreciation of this sector to obtain the gross nonwage income of this sector.

Biomass (Sector 27 of SAM) In Sect. 2.3.2, we have estimated GVAs of the sectors under biomass sector. To separate out GVA of same into wage and nonwage income component, we first decompose the GVA of each sector under biomass (like agriculture residual, firewood, animal husbandry residual, food and beverages industry residual, and paper industry residual) and then we add up their wage and nonwage income to obtain the same for biomass sector. This is estimated by using the ratio of wage and nonwage income (inclusive of depreciation) of their corresponding mother sectors like agriculture, forestry, animal husbandry, food and beverages, and paper industry sectors.

Construction (Sector 29 of SAM) The total income which is mixed of all types excluding the interest payments under unorganized sector is assumed as wage income. For organized sector, the wage and nonwage income are separately available from NAS.

Distribution of Sectorwise Consumption Expenditure by Occupational Categories of Households In the updated IO table, we have obtained the sectorwise (commodity) private final consumption expenditure (PFCE) for the year 2006–2007. Now, we have to decompose these PFCEs into the nine household classes. The NSSO in its 62nd round gives the distribution of monthly per capita household expenditure for more detailed level of classification of commodities for 2006–2007 (NSSO 2008). A map of concordance between our sectors and NSSO's item is shown in Table 2.5. On the other hand, NSSO also gives the distribution of rural population among the five rural occupational household classes and the urban population among the four urban occupational household classes for the year 2005–2006 (NSSO 2008). We use this ratio to distribute the total number of rural and urban populations into different occupational categories for the year 2006–2007. Once we have the number of population belonging to the different occupation classes, we can easily estimate the sectorwise monthly consumption expenditure of different household classes for the year 2006–2007. Thus, we have obtained the monthly consumption expenditure of the nine households group according to 28 sectors (shown in Table 2.5) of our SAM for the year 2006–2007. The output of nine sectors (like forestry, crude oil, natural gas, minerals, fertilizer, cement, iron and steel, aluminum, and construction) of our SAM 2006–2007 are not consuming by the households. The biomass consumption includes firewood consumption, so forestry output is not appearing in

Table 2.5 Mapping between social accounting matrix (SAM) sectors and National Sample Survey Organization (NSSO) items. (Source: NSSO 2008)

SAM sectors	Item code
Paddy rice	101–106
Wheat	107–114
Cereals, grains, etc., other crops	115–153
Cash crops	234, 250–257, 290, 291–293, 297, 310–330
Animal husbandry and production	160, 162–167, 180, 182–186
Fishing	181
Coal	347
Food and beverages	161, 195–286, 294–326
Textile and Leather	362–394
Wood	494
Petroleum and coal prod.	344, 345, 508, 510, 511
Chemical, rubber, and plastic production	452, 453, 456, 461, 465, 467
Paper and paper production	400, 401, 403,
Other manufacturing	440–451, 454, 455, 457, 458, 460, 462, 550–557
Machinery	560–568, 590–608, 610–614, 620, 621, 630–634, 640–643
Hydro	342
Non-hydro	342
Nuclear	342
Biomass	341, 343
Gas manufacture and distribution	348, 353
Water	540
Land transport	502, 503, 505, 506
Rail transport	501
Air transport	500
Sea transport	504
Health and medical	410–424
All other services	430–435, 402, 404–406, 480–494

the consumption flow. In this way, we estimate the itemwise share of consumption expenditure for each household group. Hence, we are in a position to derive the consumption expenditure of each of the household classes of our SAM using these shares to the PFCEs of these sectors of the year 2006–2007.

The total of indirect taxes on PFCE is divided into taxes paid by different households categories in proportion to the total expenditures of these categories on non-agricultural commodities. There is no such tax on agricultural commodities. The total of the expenditures on different sectors and the taxes paid is equal to the total expenditure for each category.

Distribution of Household Income by Source of Income and by Wage and Other Components As noted earlier, we have considered nine household classes in our SAM. So, we need to estimate the total personal income of each of these nine households' classes. In general, the households receive income from different sources like labor income, income from capital owned by households, land income, and transfer income from government and ROWs. We need to account each of them to derive personal income of each of the household class.

The households receive wage income due to supply of their labor force to the production sectors of the domestic markets as well as international markets. Earlier in this chapter, we have estimated the payment for wage for each of the domestic sectors of our SAM. On the other hand, the net wage income from ROW for the year 2006–2007 is available from NAS. We have added up these wages payments of the domestic production sectors and the net wage income from ROW to estimate households' total labor income. To distribute this labor income among the nine households' classes, we have used SAM of the year 2002–2003 constructed by Pradhan et al. (2006). In that SAM, the total labor income has been distributed among the nine households' classes for the year 2002–2003. Since the household classes of that 2002–2003 SAM correspond to our proposed household classes, we have used the share of each households labor income of the year 2002–2003 to the total labor income of the year 2006–2007.

Earlier in this chapter, we have estimated the payment of the domestic production sectors for the capital use for the year 2006–2007. This payment for capital along with net capital income from ROW is treated as gross capital income of the economy where the data on net capital income from ROW is available from NAS. We subtract the depreciation from the gross capital income to obtain the net capital income of the economy. Now this net capital income is not only received by the household classes but also by the private corporate sector, public nondepartmental enterprises, and government. The private corporate sectors receives the capital income in the form of operating profits, the public nondepartmental enterprises receives the same in the form of operating surplus, and the government receives capital income in the form of entrepreneurship income. The data on operating surplus, operating profit, and the income from entrepreneurship are available from NAS for the year 2006–2007. So the remaining part of the capital income is capital income of the households. We have distributed this capital income among the household classes by using the share of each households capital income as available from SAM of the year 2002–2003.

Next we have to estimate the land income received by the household classes. Understandably, only the rural agricultural self-employed class households receives the income from land. In this case, we have taken the total payment for land factor as the total land income of that class.

The other sources of households' income are transfer income from government and the net current transfer from the ROW. NAS gives data on current transfer from government and also the net current transfer from the ROW. Now, the government transfer includes direct government transfer to the households and the other is interest payment for debt. A part of this interest payment of the government goes to the private corporate sectors due to their holding of public shares, bonds, etc. We estimate the interest income received by the private corporate sector using information from SAM 2002–2003. The remaining part of the government transfer is distributed among the household classes by using the ratios given in SAM 2002–2003 (Pradhan et al. 2006).

Thus, we obtain the households' personal income from different sources, i.e., row total of each household class in our SAM. The households' personal income obtained in the above way did not match with the column total of each of the household classes of our SAM. A pro-rata adjustment has been made to obtain the control total, i.e., row total of each household class in our SAM (Pradhan et al. 2006).

2.3.6 Construction of Tax Account (Direct and Indirect Taxes)

Though indirect taxes are part of the government activities, we have made it a separate account in order to simplify the presentation of the detail structure of taxes. The indirect taxes reported in the SAM are net of subsidies (net indirect taxes). The net indirect taxes on household consumption and government consumption are inclusive of sale taxes and excise for domestic production, and taxes on imported commodities (custom) used for consumption. The decomposition of net indirect taxes across production sectors is done with the help of the IO table of the year 2006–2007.

Total direct taxes as obtained from the NAS are distributed among different categories of households in the following manner. Land revenue is paid by self-employed agricultural households. The other direct taxes are distributed among different categories of households in proportion to their personal income, assuming no direct tax to be paid by agricultural and nonagricultural labor households and self-employed rural agricultural households.

2.3.7 Construction of Capital Account

This account represents the macro balancing of savings and investments. Net savings include those by the households, the private corporate sector, the public non-departmental enterprises, the government, and the ROW. Net saving along with the depreciation equals gross domestic capital formation. In the case of households, the savings of different categories are derived by subtracting their consumption and direct taxes from their total personal income.

Retained earnings of the private corporate sector and the nondepartmental public enterprises are treated as their savings. The difference between the revenue and current expenditure of the government is its saving. Foreign savings meet the difference between gross domestic capital formation and gross domestic saving.

2.3.8 Treatment of Foreign Trade

The record of foreign trade is given under the row and column head of ROW in our SAM. As Table 2.1 indicates, the column of ROW describes the exports of goods and services, net factor income from abroad, net capital transfer to the government, other current transfer to the households, and private corporate and foreign savings. The data on exports are directly available from IO table 2006–2007 and the data other than exports are available from NAS and the published statistics of Reserve Bank of India (RBI 2006). We put these data under the column of ROW in such a manner that the column total of ROW must be equal to the total imports. In the IO table 2006–2007, the imports are recorded in the column of imports with negative entries. But in our SAM, we put this data under the row head of ROW with positive entry.

2.4   The SAM for India 2006–2007

The SAM computed for India consists of 35 producing sectors, 3 factors of production, and 4 economic institutions including household sector, which are classified into 9 broad categories according to their occupation. The IO table for 2006–2007 has been first expanded to match with sectors of our SAM and then we extend the various accounts (viz., GVA, PFCE, personal income, etc.) of IO table to make these as accounts of a SAM. This constructed SAM is given in Appendix 2.

The commodity X commodity part of this matrix gives the interindustry flow of intermediate inputs. The rows with heads Labor, Capital, and Land give the primary input into each sector. Only the agricultural sectors are using these three as primary inputs, and the rest of the sectors are using only labor and capital as primary inputs in their production process. Interestingly, the net factor income from abroad (under column head *ROW* and rows *Labor* and *Capital*) are seen to be substantially negative. This shows that there is a net repatriation outside the country.

The factors of production receive their factor payments, which in turn go to the different household categories as their income. The values can be read from the rows for rural and urban households and the columns under *Labor, Capital,* and *Land.* The households also receive transfer payments from the government and from ROW.

The private and public enterprises earn gross profits on account of capital and the private firms also receive net transfers from the government. The row under *Capital A/C* gives the savings. These are calculated as residuals and hence are balancing entries. The government is seen to be a net dissaver.

The imports of various commodities are given by the entries in the row corresponding to the ROW. Under the column ROW, we have exports of the commodities, net factor income from abroad, transfers to the rural and urban households, net indirect taxes, and foreign savings.

In summary, this chapter contributes to existing literature by constructing SAM of India for the year 2006–2007. This is the most updated macro database for India with detail of energy sectors. Meanwhile, it must be noted that the SAM itself is not a model but merely a database. Once the closure rule is specified, SAM becomes a model. With the help of multiplier model, one can compute income and employment effects at the sectoral level of various policies shocks. Although SAM has significantly extended the multisectoral framework of the IO table, it still falls short in representing elements such as pollutants, environmental quality, natural resources, and most of their interactions with economic activities in the real world. Neither the impact of economic activities on the environment nor the constraints of the environmental quality on production and welfare have been reflected directly in a SAM framework. In the next chapter, we will extend the SAM framework to account for pollution-related activities and which will be a unique addition to the existing literature on macro database for India.

References

Bhide S, Pohit S (1993) Forecasting and policy analysis through a CGE model for India, Margin, NCAER vol 25, no 3. New Delhi, pp 271–285

Central Statistical Organisation (1989) National accounts statistics of India: sources and methods, Government of India.

Central Statistical Organisation (2007) Energy Statistics 2007, Ministry of statistics and programme implementation, Government of India

Central Statistical Organisation (2008) Input output transaction table (2003–04), Government of India

Chowdhury A, Kirkpatrick C (1994) Development policy and planning: an introduction to models and techniques. Routledge, London

CSO (Central Statistical Organization) (2010) Input output table for India (2006–07), Ministry of statistics and program implementation, Government of India

Fritz B, Steininger K (1998) Biomass energy use to reduce climate change: a general equilibrium analysis for Austria. J Policy Model 20:513–535

National Sample Survvey Organisation (NSSO) (2008) Household consumption expenditure 2005–06, 62nd round, Government of India, Report No.523

Pal, BD, Pohit S, Roy S (2012) Social accounting matrix for India. Econ Syst Res 24(1):77–99

Pradhan BK, Sahoo A (1996) Social accounting matrix and its multipliers for India, Margin, NCAER, New Delhi, Jan–May, 28, and no.2

Pradhan BK, Sahoo A, Saluja MR (1999) A social accounting matrix for India 1994–95. Econ Polit Wkly, 34(48), 3378–3394

Pradhan BK, Saluja MR, Singh SK (2006) A social accounting matrix for India, concepts, construction and applications. Sage, New Delhi

RBI (2006) Hand book of statistics on Indian economy. Reserve Bank of India, Government of India

Round JI (1981), Income distribution within a social accounting matrix: a review of some experience in Malaysia and other L.D.C.'s, Development Economics Research Centre-University of Warwick, discussion paper no: 3, March 1981

Saluja MR, Yadav B (2006): The social accounting matrix of India. Planning Commssion, Government of India

Sarkar H, Panda M (1986) Quantity-price money interaction in a CGE model, Margin, NCAER, New Delhi, vol 18, no 3, pp 31–47

Sarkar H, Subbarao (1981) A Short term macro forecasting model for India: Structure and uses. Ind Econ Rev 16:55–80

Sinha A, Siddiqui KA, Munjal P (2007) A SAM framework for the Indian informal economy. Oxford University Press

Chapter 3
Environmentally Extended Social Accounting Matrix of India: Definition and Construction Methodology

The accounting for pollution-related activities within a social accounting matrix (SAM) framework is useful for analysis of environment-related issues. In this chapter, we describe the methodology one uses for accounting the same in a SAM framework. In other words, how to construct an environmentally extended SAM (ESAM), is spelt out in detail here. We have used this methodology to convert our constructed SAM into an ESAM. Basically, the ESAM presented in this chapter provides an integrated account for both economic transactions in monetary unit and flow of environmental substances in physical unit. Such integrated account in a consistent framework helps us to understand and quantify the linkages between economic activities and the flow of environmental substances.

3.1 The Framework and Methodology

Theoretically speaking, the extension of a SAM can be presented either in additional rows and columns, or in satellite tables (Keuning 1992). An example of an extension by means of satellite tables is the National Accounting Matrix including Environmental Accounts (NAMEA) for the Netherlands for the year 1992 (Keuning 1992). In NAMEA, the national accounting matrix of Netherlands is extended with three accounts on the environment. A substances account, an account for global environmental themes and an account for national environmental themes. These accounts do not express the transaction in money terms but include the information in physical unit.

In this study we have followed this NAMEA approach to construct the ESAM for India for the year 2006–2007. In the Table 3.1 given below, we have described the framework of ESAM for India.

As Table 3.1 shows, the first six accounts are the accounts for a SAM and details of these accounts are described in the previous chapter. The remaining accounts of Table 3.1 describe the information related to environment. The accounts that are related to environment are the substances account (account 7, 8 and 9 in Table 3.1) and the account for environmental themes (account 10 in Table 3.1). The description of these accounts is given in the following paragraphs.

B. D. Pal et al., *GHG Emissions and Economic Growth,*
India Studies in Business and Economics, DOI 10.1007/978-81-322-1943-9_3,
© Springer India 2015

Table 3.1 Schematic structure of environmentally extended social accounting matrix.
(Source: National accounting matrix for environmental accounting (NAMEA) by Keuning 1992)

			Production	Factors of production	Institutions	Indirect taxes
			1	2	3	4
Production		1	Intermediate consumption		Consumption of goods and services	
Factors of production		2	Payment for factors			
Institutions		3		Value added income	Transfer from other institutions	Total tax receive
Indirect taxes		4	Taxes on intermediate		Taxes on purchase	
Capital account		5		Depreciation	Savings	
Rest of the world		6	Imports			
Damaging substances (pollutants)		7	Absorption of substances in production		Absorption of substances in consumption	
Depletable substances	Depletion of energy resources	8		Depletion of energy stock		
	Depletion of land	9		Depletion of land through conservation		
Environmental themes	Greenhouse effect	10		Accumulation of greenhouse gases		

Capital account	Rest of the world	Substances (pollutants)	Depletion of natural resource		Environmental theme
			Renewal of energy resource	Renewal of land	Greenhouse effect
5	6	7	8	9	10
Change in stocks and capital formation	Exports	Emission of pollut-ants from production			
	Net factor income from abroad		Renewal of energy capital	Renewal of land capital	
	Net current transfers	Emission of pollut-ants from consumption			
Taxes on investment					
	Foreign savings				
			Removal of sub-stances	Accumulation of substances	
				Reduction in natu-ral stock	
		Emission from land use change			

The "substances account" of the ESAM provides flow data on the supply and use of a number of substances that affect in one way or the other the natural assets like air, water, etc., and create pollution (Keuning 1992). The term "substances" refers not only the matter which is of damaging in nature (e.g., emissions of chemicals, wastes, etc.) but also includes valuable matter in the form of depletable natural resources. In case of damaging substances, the columns of the substances account describe only the supply of these substances from different source and the rows describe the absorption of these substances into different sectors. But in case of depletable substances, the column shows the renewal of the natural capital and this comes from the new discoveries of exhaustive natural assets like coal, crude oil, etc. On the other hand, the row of this depletable substances account shows the use of these substances in the form of intermediate input in the production process. Finally, the account for environmental themes is an inventory account which takes into account the net emission (i.e., emission minus absorption) in India. In this way, we have described all the accounts of our proposed ESAM for India. Of course, we have to collate the relevant data for the construction of ESAM, which is described in the following sections.

3.2   Estimation of Environmental Data

The substances accounts and the account for environmental themes are generally expressed in physical unit in an ESAM. So, we have to estimate the environmental data in physical unit. Since our ESAM is an extension of our SAM, we have to estimate the environmental data for 35 production and consumption activity of nine household groups for the year 2006–2007.

3.2.1   Estimation of Generation of Damaging Substances

There are different types of damaging substances that exist in the environment which causes damage to natural resources in the form of air pollution, water pollution, etc. But, our concern in this study is about the substances which are responsible for climate change. Therefore, we have to estimate the supply and absorption of the greenhouse gases (GHGs) such as CO_2, CH_4, and N_2O in India to construct the column and row for the damaging substances for this ESAM for 2006–2007.

The GHG emissions in India are observed from the domestic production process, import, and from consumption activities (TEDDY 2009). But in India the data on GHG emissions through import are not available. So we have taken only the production and consumption-based supply of these damaging substances in India.

India has published its second communication on greenhouse gas emission, (Indian Network on Climate Change Assessment, INCCA 2010) which provides updated information on India's GHG emissions from different sectors for the year

Table 3.2 Mapping between environmentally extended social accounting matrix (ESAM) sector and sector of Indian Network on Climate Change Assessment (INCCA) report. (Source: authors' estimate)

Sector of ESAM	INCCA sectors
Paddy rice	Agriculture, rice cultivation, soils
Wheat	Agriculture, soils
Cereals	Agriculture, soils
Cash crops	Agriculture, soils
Animal husbandry	Agriculture, enteric fermentation, manure management
Forestry	No emission
Fishing	Agriculture
Coal	Fugitive emission
Oil	Fugitive emission
Gas	Fugitive emission
Food and beverages	Food processing, industrial wastewater
Textiles and leather	Textile and leather, industrial wastewater
Wood	Nonspecific industries
Minerals not elsewhere classified	Mining and quarrying
Petroleum and coal tar product	Other energy industry, industrial wastewater
Chemical, rubber, and plastic products	Chemicals, industrial wastewater
Paper and paper products	Pulp and paper, industrial wastewater
Fertilizer and pesticides	Nonspecific industries, industrial wastewater
Cement	Cement
Iron and steel	Iron and steel
Aluminum	Aluminum
Other manufacturing	Ferroalloys, lead, zinc, copper, glass and ceramic, soda ash, nonspecific industries
Machinery	Nonspecific industries, industrial wastewater
Thermal	Electricity, industrial wastewater
Hydro	No emission
Nuclear	No emission
Biomass	No emission
Water	No emission
Construction	Nonspecific industries
Land transport	Road transport
Railway transport	Railways
Air transport	Aviation
Water transport	Navigation
Health and medical	No emission
Other services	Commercial/institutional

2007 (INCCA 2010). In this study, we have taken this report as base to estimate sector-specific GHG emissions of India for the year 2007. However, the sector described in INCCA report does not exactly match with the sectors of our SAM of the year 2006–2007. So to estimate sectorwise GHG emissions, we have made a map of concordance between our SAM sectors and the sectors given in INCCA report which is given below (Table 3.2).

As this table shows, most of sectors of our ESAM do not have one-to-one correspondence with the INCCA sectors. So, we have to disaggregate the GHG emissions for the INCCA sectors for which there is no one-to-one correspondence with our ESAM sectors. The method of desegregation is described below.

3.2.1.1   Decomposition of Agricultural Sector's GHG Emissions

According to INCCA report, the GHG emissions from agriculture sector arises from energy use and production process. The GHGs emitted from energy use in agriculture sector are CO_2, CH_4, and N_2O, whereas CH_4 and N_2O are emitted from production process. Except forestry, all the agricultural sectors use energy for their energy requirement and so the energy-based GHG emissions will be observed for these sectors. By and large, the rice cultivation sector and animal husbandry sector emit CH_4 through their production process. On the other hand, N_2O emission from agricultural production process is observed due to fertilizer use into the soils and it is also observed for all agricultural sectors except forestry.

In our SAM, there are six agricultural sectors barring forestry among which rice cultivation and animal husbandry sector have separate account. So, CH_4 emissions from the production process of rice cultivation and animal husbandry are directly obtained from the INCCA report. But to obtain the energy-based emission from the agricultural sectors of our ESAM, we have to disaggregate the energy-based GHG emissions from agricultural sector into these six agricultural sectors of our ESAM. In this case, we have estimated share of energy use of these six sectors from our ESAM and apply this to obtain their energy-based GHG emissions for these sectors.

Next, we have disaggregated the N_2O emission from soils into the six agricultural sectors of our SAM and this is done on the basis of the share of fertilizer use of these sectors. The share of fertilizer use for these sectors is obtained from our SAM of the year 2006–2007. Thus we have obtained GHG emissions from all the agricultural sectors of our SAM.

3.2.1.2   Desegregation of Fugitive Emissions

CH_4 emission from this source is only observed during extraction, production, process, and transportation of fossil fuels such as coal, crude oil, and natural gas (INCCA 2010). CH_4 emission from this source is 730,000 tons from coal mining and 779,400 tons from combined crude oil and natural gas sector. Therefore, we have to estimate emission from this source separately for crude oil and natural gas sector and this is done on the basis of INCCA report. INCCA report provides us the emission coefficients of CH_4 from this source separately for crude oil and natural gas sector. To apply these coefficients, we have taken production data of crude oil and natural gas sector from energy statistics of India (CSO 2007). Thus, we have estimated CH_4 emission from this source for coal, crude oil, and natural gas sector separately for India for the year 2007.

Table 3.3 Wastewater generation in Indian industry. (Source: INCCA report 2010)

Sector	Wastewater (million L/day
PAP (Paper and paper products)	1881
PET (Petroleum and coal tar products)	143
FER (Fertilizer)	168
IRS (Iron and steel)	1087
NHY (Thermal electricity)	72,219
FBV (Food and beverages)	424
TXL (Textiles and leather)	1930
CHM (Chemicals)	213

3.2.1.3 Desegregation of Nonspecific Industries GHG Emissions

In this case, we have first estimated the energy-based emission for the sectors of our SAM with nonspecific industries mentioned in INCCA report. We have primarily used energy use data from annual survey of industries (ASI; CSO 2006–2007) for the estimation exercise. The data obtained from ASI gives energy use in value terms. To convert these into physical unit, we have estimated general price level (value of energy/quantity of energy) on the basis of energy statistics of India (CSO 2007). We have also applied energy-specific emission coefficients given in Intergovernmental Panel on Climate Change (IPCC) guidelines (www.ipcc.ch) to obtain GHG emissions from these sectors. The GHGs emissions obtained in this way do not match with the GHG emissions given in INCCA for this sector. So, we have considered the INCCA estimate as control total and a pro rata adjustment method has been applied to get this control total.

3.2.1.4 Desegregation of Industrial Wastewater

The INCCA reports CH_4 emission from industrial wastewater also. Since there is no separate account of this sector in our ESAM, we have added this emission to the CH_4 emission of the industries-generated wastewater. In the following table, we have provided the industrywise generation of wastewater (Table 3.3).

Now to estimate CH_4 emission from these sectors, we have disaggregated the total emission from industrial wastewater into these sectors. To disaggregate this emission, we have used share of the above industries in total industrial wastewater generation.

3.2.1.5 Estimation of GHG Emissions from Household Activity

Households are categorized into nine types of household classes in our SAM. So, we have to estimate GHG emissions from household activities for these nine types of households. To estimate the GHG emissions from household activities, we have used the following estimates of INCCA:

1. Estimated GHG emissions from residential source
2. Estimated GHG emissions from burning of crop residue
3. Estimated GHG emissions from fuel wood use
4. Estimated GHG from municipal solid waste
5. Estimated GHG emissions from domestic wastewater

The above estimates give aggregate GHG emissions for all households. To disaggregate these into five household classes, we have adopted the following approach.

Among the above five estimates, the emissions from residential source is based on emissions from cooking coal, liquid petroleum gas (LPG), and kerosene (INCCA 2010). To disaggregate GHG emissions from this source, we have used share of energy consumption (excluding biomass) of each household class as obtained from our SAM. Thus, we have estimated GHG emissions from residential sources for each household classes of our SAM.

The GHG emissions from burning of crop residue and fuel wood use are highest for the rural households in India (INCCA 2010). We have to disaggregate GHG emissions from these two sources among the rural household classes, and this is done on the basis of share of biomass consumption of the rural households.

On the other hand, the urban households are the sources of emissions from municipal solid waste and domestic wastewater (INCCA 2010). Logically, we need to disaggregate these emissions into the urban household classes of our SAM. INCCA reports that the urban households generate 0.55 kg/capita/day of municipal solid waste in India. We have used this data to estimate solid waste generation for each of these household classes of urban area. Once we obtained the solid waste generation for each of the urban household classes, we can estimate the share of waste generation for each household classes of urban area. Using this share, we have estimated the household-classwise GHG emissions from municipal solid waste in India for the year 2006–2007.

We also need to estimate the GHG emissions from domestic wastewater for each of the urban household classes for the year 2006–2007. The share of consumption expenditure on water for the urban household classes of our SAM has been used to estimate the household-classwise GHG emissions from this source for urban area.

3.2.1.6 Emission Due to Land Use Change

The changes in grassland resulted in the emission of 10.49 million tons of CO_2 due to decrease in grassland area by 3.4 million hectare between 2006 and 2007 (INCCA 2010).

Hence, in the abovementioned way, we have estimated generation of GHGs from production as well as consumption activities for India for the year 2006–2007.

3.2.2 Estimation of Abatement or Absorption

In India, the data on GHG abatement are not available for production and consumption activities. It is observed from the INCCA report that the Land Use Change and Forestry (LULUCF) play major role in CO_2 removals in India. The gross CO_2 removal from LULUCF is 27.53 million tons in the year 2006–2007 (INCCA 2010). In this study, we have used this data to construct the row of damaging substances of our proposed ESAM for India.

3.2.3 Estimation for Depletable Natural Resources

Here we consider crude oil, coal, and land as the depletable natural resources in our ESAM. The production data in physical unit on crude oil and coal have been taken as measures of the quantities of depletion of these two types of resources. The source of data is various issues of energy statistics by CSO (CSO 2006, 2007). The data obtained in this way can be interpreted as "free" intermediate consumption (without direct cost) used in the production process of the crude oil and natural gas sector. To construct the row of this depletable substances account, we have put the data in the row of this depletable substances account corresponding to the column of crude oil and coal sectors.

To construct the column of this depletable substances account, we have obtained the data on new discoveries of crude oil and coal reserve in India in the year 2006–2007. The source of our data is TERI (2009).

In this way, we have constructed the row and the column of the depletable substances account of our ESAM for India for the year 2006–2007.

We have used data from INCCA report to construct the row of land use change. INCCA report gives change in different types of land use data for the year 2006 and 2007. The difference between these 2 years' data is used as land use change in the year 2007. The same report also gives the data on land conservation in India during 2006 and 2007. We have used this data to get the column of the land use change under the item head of "renewal of natural resources" in our proposed ESAM. Thus, we have obtained the row and column for each of the substances to construct our ESAM for India for the year 2006–2007.

Apart from this substances account, there is also another account, i.e., account for environmental themes which is also important for our ESAM for India. Below, we have described the method of estimating environmental themes for India for the base year of our ESAM.

3.2.4 Estimation for Environmental Themes

In this case, we have to estimate the inventory of the GHGs in the Indian economy for the year 2006–2007 to show greenhouse effect. To estimate this greenhouse effect, we have to estimate the net generation of GHGs in terms of carbon dioxide equiva-

Table 3.4 Direct pollution coefficient matrix 2006–2007 (unit tons/lakh of output). (Source: authors' estimate)

Sector of ESAM	CO_2	CH_4	N_2O
PAD	0.77	0.22	0.00
WHT	0.84	0.00	0.00
CER	0.55	0.00	0.00
CAS	0.28	0.00	0.00
ANH	2.31	0.48	0.00
FRS	0.00	0.00	0.00
FSH	0.00	0.00	0.00
COL	0.00	0.13	0.00
OIL	0.00	0.00	0.00
GAS	0.00	0.40	0.00
FBV(Food and beverages)	0.74	0.00	0.00
TEX	0.07	0.00	0.00
WOD	0.30	0.00	0.00
MIN	0.12	0.00	0.00
PET (Petroleum and coal tar products)	1.03	0.00	0.00
CHM (Chemicals)	0.53	0.00	0.00
PAP (Paper and paper products)	1.31	0.01	0.00
FER (Fertilizer)	4.85	0.00	0.00
CEM	34.67	0.00	0.00
IRS (Iron and steel)	4.00	0.00	0.00
ALU	0.21	0.00	0.00
OMN	0.54	0.00	0.00
MCH	0.26	0.00	0.00
NHY (Thermal electricity)	47.83	0.07	0.00
HYD	0.00	0.00	0.00
NUC	0.00	0.00	0.00
BIO	0.00	0.00	0.00
WAT	0.00	0.00	0.00
CON	0.00	0.00	0.00
LTR	2.37	0.00	0.00
RLY	0.70	0.00	0.00
AIR	9.61	0.00	0.00
SEA	0.83	0.00	0.00
HLM	0.00	0.00	0.00
SER	0.01	0.00	0.00

lent (CO_2EQ) for the same year. To estimate generation in terms of CO_2EQ, we have to multiply CH_4 emission by 21 and N_2O emission by 310 (INCCA 2010). Thus, we have estimated environmental themes for Indian economy for the year 2006–2007.

Hence, with the abovementioned way, we have estimated ESAM for India for the year 2006–2007. This complete ESAM is given in the Appendix 3. To our best knowledge, this is the first ESAM for India which integrates the economic as well as environmental indicators in a single accounting framework. It would be interesting to analyze the relationship between the environment and economic activities for India using the structure of our ESAM. As a first step, we have estimated the direct pollution coefficient directly from the ESAM data (Table 3.4). The direct pollution coefficient is nothing but the amount of pollution generated for each level of output produced by the production activity. This direct pollution coefficient also measures the direct effect of economic activities on the environmental pollution.

The data in Table 3.4 indicates that the direct pollution coefficients are higher for the sectors like Thermal electricity (NHY), CEM, and iron and steel (IRS) sector. It implies that these three sectors have large direct impact on the environmental pollution in comparison to the other production sectors of the Indian economy. According to this table, NHY sector has the highest direct pollution coefficient for each type of pollutants compared to the other sectors. In other words, NHY sector has the highest direct effect on CO_2 emission in India. To clarify, NHY sector generates 47.83 tons of CO_2 to produce INR 1 lakh value of its output. Thus, we can understand the direct effect of the economic activities on the environmental pollution with the help of direct pollution coefficients.

In summary, the ESAM of the year 2006–2007 prepared and presented in this chapter shows the interaction between all the economic agents and their contribution in GHG emissions in India. Since the basis of an ESAM is SAM and IO table, the assumption of excess capacity assumption of IO analysis holds true for ESAM also (Pradhan et al. 2006). Therefore, if output of a sector increases due to any exogenous reason, the output of the other sector will increase. Given the technological condition, this increase in output will increase GHG emissions of the economy and this can be observed from the positive value of the direct emission coefficients as described in the Table 3.4. Apart from this direct impact, there are indirect and induced impacts on the economy whenever there are any exogenous changes in the economy. The indirect effects are the secondary effects or production changes in industries with backward linkage caused when inputs needs change due to the impact of directly affected industry. The induced effects represent the response by all industries caused by increased expenditures of new household income and interinstitutional transfers generated from the direct and indirect effects of the change in final demand for a specific industry. Therefore, any exogenous changes cause direct, indirect, and induced impact on GHG emissions. Hence, it is clear that any economic activity will have impact on the GHG emissions of the economy.

However, at what extent the GHG emissions will increase and which sector is more responsible for this is a policy-relevant question for the Indian economy. To answer this question, we assess empirically the impact of economic growth on GHG emissions for India and this is described in the following chapter.

References

Central Statistical Organisation (2006) National account statistics. Government of India

Central Statistical Organisation (2007) Energy statistics 2007. Ministry of statistics and programme implementation, Government of India

INCCA (Indian Network on Climate Change Assesment) (2010) India: greenhouse gas emission 2007. Ministry of Environment and Forests. Government of India

Keuning SJ (1992) National accounts and environment; the case for a system. Occasional paper
    Nr, NA-053, Statics Netherlands, Voorburg
Pradhan BK, Saluja MR, Singh SK (2006) A social accounting matrix for India, concepts,
    construction and applications. Sage, New Delhi
TEDDY (2009) The energy data directory year book. The energy and resource institute, New Delhi
TERI (2009) (The Energy and Resource Institute). The energy data directory of India (TEDDY),
    TERI, New Delhi

Chapter 4
Impact of Economic Growth on Greenhouse Gas (GHG) Emissions—Social Accounting Matrix (SAM) Multiplier Analysis

In general, direct pollution effect is only a small part of the total effect of pollution when production/consumption activities take place in an economy. To estimate the same, we have to undertake multiplier analysis. In this chapter, we have followed the social accounting matrix (SAM) multiplier method to estimate empirically the multiplier impact of economic activity on greenhouse gas (GHG) emissions in India. To be specific, this chapter explains the linkage of output growth with energy demand and GHG emissions. Finally, the issue of green employment opportunity has been analyzed with the help of SAM multiplier.

4.1 Approach to Link Economic Growth with GHG Emissions

Achieving economic growth, alleviating poverty by creating employment opportunity, and reducing GHG emissions are key policy issues for the late-industrializing economy like India. In this context, since energy demand is crucial for GHG emissions, the Government of India has taken various policy measures in its 12th 5-year plan (2012–2017) to improve energy efficiency in the industries (Planning Commission 2011). Again the SAM of the year 2006–2007 presented in Chap. 2 shows that the energy demand pattern is different for different sectors. Hence, the impact of growth in economic activities will have different impact on energy demand and hence GHG emissions.

Since then, a SAM multiplier describes growth in economic activity due to exogenous policy shock into the economy, linking SAM multiplier with the sector-specific energy and GHG emissions intensities; one can estimate the multiplier impact of economic growth on energy use and GHG emissions. Like a SAM multiplier, this multiplier impact will also include the direct, indirect, and induced impact of economic growth on GHG emissions while there is any exogenous change in the economy (Robert 1975).

B. D. Pal et al., *GHG Emissions and Economic Growth,*
India Studies in Business and Economics, DOI 10.1007/978-81-322-1943-9_4,
© Springer India 2015

In this study, we have assumed Government Account, Account of Net Indirect taxes, and Foreign Trade accounts of our SAM as exogenous. Keeping these accounts as exogenous, we have done the following impact analysis:

1. The impact on energy use due to growth in sectoral output resulting from any exogenous changes in the economy
2. The impact on GHG emissions due to growth in sectoral output resulting from any exogenous changes in the economy

Apart from these linkages between output growth, energy consumption, and GHG emissions, linking GHG emissions with employment generation is also crucial from the point of view of green growth. Recently, the concept of green employment is emerging across the globe and researchers are trying to find out option for green employment creation as an outcome of global climate change mitigation action (United Nations Environment Program, http://www.unep.org/civil-society/Implementation/GreenJobs/tabid/104810/Default.aspx). Since unemployment is like a chronic disease in the late-industrializing economies, policy thrust may go towards more employment generation. In this context, identifying sectors which have large impact on employment and less impact on GHG emissions would be a crucial policy challenge for the late-industrializing country like India. Therefore, understanding linkages between employment and GHG emissions is prerequisite for this analysis.

Now the changes in the above-mentioned exogenous variable may increase the employment opportunities within the domestic economy. Expectedly, pattern the employment opportunity will not increase uniformly across the sectors due to differences in labor intensity. To what extent the sectorwise employment will change is an important policy-relevant question. To answer this question, we have also derived employment multiplier from our SAM. The employment multiplier will show how the sectorwise employment changes due to any exogenous changes into the economy. So if we link this employment multiplier with the GHG emissions, we can show the impact of employment change on GHG emissions.

4.2   Method of Estimating SAM Multiplier

Before we estimate SAM multiplier, let us describe a SAM in a simple form for the understanding of multiplier analysis (Table 4.1).

The four main sets of accounts, viz., factors, institutions, activities, and capital accounts are endogenous. All other accounts are collected together as fifth block in the schematic presentation in Table 4.1 and this is also exogenous account. They include the accounts for government account, indirect taxes, and combined current and capital accounts for rest of the world. In Table 4.1, T_{13} is a matrix of value added generated by the various production sectors, T_{33} gives the input–output transaction matrix. T_{21} maps the factor income distribution into the households income distribution and T_{32} reflects the expenditure pattern of the various institutions.

Table 4.1 Schematic structure of social accounting matrix (SAM). (Source: Pradhan et al. (2006), concept and construction of SAM)

		Expenditure					
		Factors	Institutions	Activities	Capital	Other A/C	Total
Receipts	Factors			T_{13}		T_{15}	Y_1
	Institutions	T_{21}				T_{25}	Y_2
	Activities		T_{32}	T_{33}	T_{34}	T_{35}	Y_3
	Capital		T_{42}			T_{45}	Y_4
	Other A/C		T_{52}	T_{53}			Y_5
	Total	Y_1	Y_2	Y_3	Y_4	Y_5	

Finally, T_{34} displays the investment demand for production activity while T_{42} shows the savings of the institutions. The later is also referred as leakages from the system.

The first step is to obtain matrices of coefficients of expenditure A_{ij} by dividing each element in the ij by corresponding sum of the column vector Y_{j}.

$$A_{ij} = T_{ij} Y_j^{-1} \ldots\ldots = 1,2,3,4 \tag{4.1}$$

Doing so implies that the accounting constraints across rows can be expressed as

$$
\begin{bmatrix} Y_1 \\ Y_2 \\ Y_3 \\ Y_4 \\ Y_5 \end{bmatrix} =
\begin{bmatrix} 0 & 0 & A_{13} & 0 \\ A_{21} & 0 & 0 & 0 \\ 0 & A_{32} & A_{33} & A_{34} \\ 0 & A_{42} & 0 & 0 \\ 0 & A_{52} & A_{53} & 0 \end{bmatrix}
\begin{bmatrix} Y_1 \\ Y_2 \\ Y_3 \\ Y_4 \end{bmatrix} +
\begin{bmatrix} X_1 \\ X_2 \\ X_3 \\ X_4 \\ X_5 \end{bmatrix} \tag{4.2}
$$

Where X_i is the row sums of sub-matrix $T_{i,5}$ for each $i = 1, 2, 3, 4, 5$. This is just another way of representing Table 4.1.

Subsequently, analysis assumes that each of the X_i's is an exogenous set of numbers and that each of the A_{ij} matrices in equation has constant elements. Combining these two sets of assumptions implies that the values of Y_1 to Y_4 can always be obtained from any assumed values of X_1 to X_4 and can be expressed as

$$
\begin{bmatrix} Y_1 \\ Y_2 \\ Y_3 \\ Y_4 \end{bmatrix} =
\begin{bmatrix} 0 & 0 & A_{13} & 0 \\ A_{21} & 0 & 0 & 0 \\ 0 & A_{32} & A_{33} & A_{34} \\ 0 & A_{42} & 0 & 0 \end{bmatrix}
\begin{bmatrix} Y_1 \\ Y_2 \\ Y_3 \\ Y_4 \end{bmatrix} +
\begin{bmatrix} X_1 \\ X_2 \\ X_3 \\ X_4 \end{bmatrix} \tag{4.3}
$$

And

$$Y_5 = A_{52} Y_2 + A_{53} Y_3 + X_5 \tag{4.4}$$

The fifth set of accounts can be derived once Y_2 and Y_3 are known. That is, once the first three of accounts are balanced. The residual balance equation is not of further interest to us. Hence, our main analysis will be Eq. 4.3. This can be written as

$$Y = AY + X \qquad (4.5)$$

$$Y = (I - A)^{-1} X \qquad (4.6)$$

This shows that Y (i.e., Y_1, Y_2, Y_3, and Y_4) can be derived from X (i.e., X_1, X_2, X_3, and X_4) through a generalized inverse $(I-A)^{-1}$. This is analogous to that of conventional input–output analysis, which is concerned with the determination of Y_3, the production activity accounts only. It is apparent that

$$Y_3 = (I - A_{33})^{-1}(A_{32}Y_2 + A_{34}Y_4 + X) \qquad (4.7)$$

Equation 4.7 is a part of our system and, therefore, completely consistent with it. It is also the end of the story in the simplest form of input–output analysis since the latter assumes that $A_{32}Y_2$ and $A_{34}Y_4$ are exogenous. Thus, in this approach, Y_3 (the level and structure of output) is derived through the inverse $(I-A_{33})^{-1}$ of direct and indirect commodity requirements on the basis of assumed demand on activities from other accounts. In Eq. 4.7, an obvious extension is to decompose these assumed demands and to allow for parts of them, $A_{32}Y_2$ and $A_{34}Y_4$ to be determined simultaneously with Y_3. The later depends on Y_2, i.e., on the level and distribution of income across institutions and on Y_4, the saving investment balance.

Sometimes, the generalized inverse $(I-A)^{-1}$ is broken down into three matrices to reflect the different mechanism at work within it, resulting from the interconnections within the system. That is, the total multiplier of SAM is written as

$$(I - A)^{-1} = M = M_3 M_2 M_1 \qquad (4.8)$$

Where, the notation M is used to indicate a multiplier matrix. Note that M has rows and columns as factors (labor, capital), institutions (private, households, etc.), activities, and capital account. In this study, we have computed the M matrix on the basis of our 35 sector SAM of the year 2006–2007. This computed M matrix is shown in the Appendix 4.

The aggregate multiplier matrix, M, shows how an increase in any element of X_i will increase the corresponding element of Y_i by at least the same amount, and may have also indirect effects on other elements. The right-hand side of the equation shows that these aggregate multipliers can be decomposed into three separate multipliers. M_1 and M_2 are called as "own effects multipliers," as opposed to M_3 which collects together cross effects. The differences between M_1 and M_2 are as follows. A change in an element in the vector X_i of X will influence Y_i for two sets of reasons.

One is that there may be transfers within the ith set of accounts so that, for example, an increase in demand on a production sector will cause it to increase its demand on another production sector. A second example is that increased income of companies from an exogenous source will result in increased income for government through profit taxation. Such multiplier processes which operate within a set of accounts can be referred to as "own direct effects" or "own transfer effects." These contrast with M_2 which collects together "own indirect effects" which arise from the fact that an increase in the elements of X_i will affect Y_i via other accounts. Thus, an increase in demand on a production activity will cause it to hire more factors. This will raise incomes in the factor accounts which in turn raise incomes in the institution accounts. These latter accounts will spend some of the increased income and in doing so will raise demand on the production account beyond the level of the initial increase which came from an exogenous source. Finally, multipliers M_3 record cross effects, i.e., the impact of an increase in elements of X_i on Y_j for $j \neq i$.

Though one can disaggregate the effects into three separate components, one needs for all practical purposes the total impact which can be obtained straightforward by working with M matrix. Thus, if our focus is on the impact on production activities due to increased spending by the household, one needs to look at the components of the matrix Mactivity × households.

So far, our discussion has said nothing about the consequence on environment of change through changes in exogenous sectors to the economy. This can be easily remedied by making assumption about the links between gross output and pollution generation in each production activity. The standard practice in this regard is to estimate direct pollution generation coefficients for each sector. These direct pollution generation coefficients are obtained from our ESAM by dividing column of each type of pollutants with the column of gross output (Table 3.4 of Chap. 3). The pollution generation coefficients thus obtained are expressed in terms of tons of pollution per lakhs of rupees of output. Once we have these coefficients, we can perform environmental impact analysis by estimating the pollution trade-off multipliers.

4.3   Pollution Trade-Off Multiplier

To estimate pollution trade-off multiplier, we have used method described by Robert (1975). The pollution trade-off multiplier measures the direct, indirect, and induced impact on pollution generation level due to exogenous change in sectoral output, households income, etc. The mathematical expression of the pollution trade-off multiplier is given as follows:

$$E = P \cdot Y \tag{4.9}$$

where E is matrix of sectorwise emission, P is the sectorwise emission coefficient matrix.

Replacing Eq. 4.6 into Eq. 4.9,

$$E = P \cdot (I - A)^{-1} \cdot X \qquad\qquad (4.10)$$

$$\frac{\partial E}{\partial X} = P \cdot (I - A)^{-1} = T \qquad\qquad (4.11)$$

T is the pollution trade-off multiplier matrix which indicates impact on emission due to any exogenous changes into the economy. T matrix has different blocks like pollutant × activity block, pollutant × households block, etc. and these can be used for different types of impact analysis. For example, if we want to analyze impact of output growth on emission we have to use pollutants × activity block of the T matrix.

We have illustrated the methodology for estimating SAM multiplier as well as pollution trade-off multiplier. In what follows, we have used our ESAM to derive quantitative estimates which are relevant in the context of the objectives of this chapter.

4.4 Impact of Sectoral Output Growth Resulting from any Exogenous Changes in the Economy on Energy Use

Any production process consumes energy. So, any output increase would lead to incremental energy consumption unless there is any innovation in energy efficient technology. But, the innovation of energy efficient technology is not a quicker process. So, our objective in this study is to analyze the sectoral growth impact on energy use under the existing technological condition. Now this SAM multiplier model is a fixed technology static model. Therefore, the SAM multiplier model does not consider any technological improvement in the economy. Rather, it takes into account the direct, indirect, and induced effect on energy use under constant technological condition.

To analyze the total effect (i.e., direct, indirect-induced) on energy use, we have considered the energy × activity part of the SAM multiplier matrix. Our SAM takes into account seven types of energy commodities. However, we are concerned here mainly with the commercial energy available from primary conventional sources. The commercial energy commodities available from primary conventional sources are coal, crude oil, natural gas, hydro, and nuclear. Table 4.2 shows the impact of sectoral growth on this primary energy use due to any exogenous changes in the economy.

In this table, we have shown the direct and indirect-induced effect on energy use for all sectors except the above four types of primary energy sectors. We have also decomposed the total effect, i.e., direct, indirect-induced effect into two parts, viz., direct and indirect-induced effects. The direct effect on energy use is nothing but the direct energy coefficient, i.e., amount of energy to be used to produce one unit

Table 4.2 Impact of sectoral growth on primary energy use for the production. (Source: Author's estimation)

Sectors	Coal			Crude oil			Natural gas			Hydro			Nuclear		
	Total effect	Direct effect	Indirect-induced effect	Total effect	Direct effect	Indirect-induced effect	Total effect	Direct effect	Indirect-induced effect	Total effect	Direct effect	Indirect-induced effect	Total effect	Direct effect	Indirect-induced effect
PAD	0.0597	0.0000	0.0597	0.2242	0.0000	0.2242	0.0267	0.0000	0.0267	0.0358	0.0038	0.0320	0.0054	0.0006	0.0049
WHT	0.0613	0.0000	0.0613	0.2197	0.0000	0.2197	0.0299	0.0000	0.0299	0.0386	0.0054	0.0332	0.0059	0.0008	0.0050
CER	0.0559	0.0000	0.0559	0.2014	0.0000	0.2014	0.0207	0.0000	0.0207	0.0302	0.0009	0.0293	0.0046	0.0001	0.0044
CAS	0.0556	0.0000	0.0556	0.2042	0.0000	0.2042	0.0223	0.0000	0.0223	0.0296	0.0007	0.0290	0.0045	0.0001	0.0044
ANH	0.0569	0.0000	0.0569	0.1948	0.0001	0.1946	0.0181	0.0000	0.0181	0.0296	0.0000	0.0296	0.0045	0.0000	0.0045
FRS	0.0211	0.0000	0.0211	0.0733	0.0000	0.0733	0.0065	0.0000	0.0065	0.0109	0.0000	0.0108	0.0016	0.0000	0.0016
FSH	0.0559	0.0000	0.0559	0.2109	0.0000	0.2109	0.0172	0.0000	0.0172	0.0287	0.0000	0.0287	0.0044	0.0000	0.0044
FBV	0.0531	0.0008	0.0523	0.1898	0.0000	0.1898	0.0179	0.0001	0.0179	0.0287	0.0011	0.0276	0.0044	0.0002	0.0042
TEX	0.0572	0.0010	0.0562	0.1968	0.0000	0.1968	0.0193	0.0013	0.0180	0.0337	0.0040	0.0296	0.0051	0.0006	0.0045
WOD	0.0578	0.0047	0.0531	0.1789	0.0003	0.1786	0.0167	0.0002	0.0165	0.0291	0.0021	0.0270	0.0044	0.0003	0.0041
MIN	0.0210	0.0020	0.0190	0.0619	0.0000	0.0619	0.0058	0.0001	0.0057	0.0105	0.0010	0.0095	0.0016	0.0001	0.0014
PET	0.0344	0.0077	0.0267	0.6704	0.5674	0.1030	0.0080	0.0000	0.0079	0.0156	0.0017	0.0139	0.0024	0.0003	0.0021
CHM	0.0494	0.0027	0.0467	0.1676	0.0041	0.1636	0.0226	0.0063	0.0163	0.0269	0.0028	0.0241	0.0041	0.0004	0.0037
PAP	0.0611	0.0133	0.0478	0.1589	0.0001	0.1588	0.0148	0.0005	0.0143	0.0287	0.0045	0.0242	0.0044	0.0007	0.0037
FER	0.0521	0.0050	0.0471	0.2284	0.0000	0.2284	0.0795	0.0586	0.0209	0.0266	0.0023	0.0243	0.0040	0.0003	0.0037
CEM	0.1103	0.0579	0.0523	0.1620	0.0000	0.1620	0.0242	0.0096	0.0146	0.0391	0.0123	0.0268	0.0059	0.0019	0.0041
IRS	0.1345	0.0706	0.0639	0.1589	0.0007	0.1582	0.0266	0.0109	0.0157	0.0338	0.0070	0.0268	0.0051	0.0011	0.0041
ALU	0.0610	0.0350	0.0261	0.0731	0.0001	0.0730	0.0088	0.0020	0.0068	0.0145	0.0027	0.0118	0.0022	0.0004	0.0018
OMN	0.0452	0.0075	0.0378	0.1055	0.0012	0.1043	0.0111	0.0009	0.0101	0.0188	0.0021	0.0166	0.0028	0.0003	0.0025
MCH	0.0561	0.0017	0.0544	0.1362	0.0001	0.1361	0.0146	0.0005	0.0141	0.0242	0.0017	0.0225	0.0037	0.0003	0.0034
NHY	0.2098	0.1131	0.0967	0.2539	0.0027	0.2512	0.0452	0.0218	0.0234	0.0996	0.0523	0.0473	0.0151	0.0079	0.0072
BIO	0.0571	0.0000	0.0570	0.1975	0.0000	0.1974	0.0179	0.0000	0.0179	0.0295	0.0001	0.0294	0.0045	0.0000	0.0045
WAT	0.0620	0.0001	0.0619	0.1944	0.0006	0.1939	0.0186	0.0001	0.0184	0.0355	0.0038	0.0318	0.0054	0.0006	0.0048
CON	0.0663	0.0000	0.0663	0.1961	0.0000	0.1961	0.0177	0.0000	0.0177	0.0307	0.0018	0.0288	0.0047	0.0003	0.0044
LTR	0.0502	0.0000	0.0502	0.3038	0.0000	0.3038	0.0151	0.0000	0.0151	0.0255	0.0001	0.0254	0.0039	0.0000	0.0039

Table 4.2 (continued)

Sec-tors	Coal			Crude oil			Natural gas			Hydro			Nuclear		
	Total effect	Direct effect	Indirect-induced effect	Total effect	Direct effect	Indirect-induced effect	Total effect	Direct effect	Indirect-induced effect	Total effect	Direct effect	Indirect-induced effect	Total effect	Direct effect	Indirect-induced effect
RLY	0.0691	0.0005	0.0686	0.2016	0.0000	0.2016	0.0188	0.0000	0.0188	0.0494	0.0147	0.0347	0.0075	0.0022	0.0053
AIR	0.0533	0.0000	0.0533	0.2323	0.0000	0.2323	0.0168	0.0000	0.0168	0.0275	0.0004	0.0270	0.0042	0.0001	0.0041
SEA	0.0554	0.0000	0.0554	0.2016	0.0000	0.2016	0.0176	0.0000	0.0176	0.0290	0.0006	0.0285	0.0044	0.0001	0.0043
HLM	0.0556	0.0000	0.0556	0.1919	0.0000	0.1919	0.0186	0.0000	0.0186	0.0292	0.0002	0.0290	0.0044	0.0000	0.0044
SER	0.0557	0.0001	0.0556	0.1864	0.0001	0.1863	0.0171	0.0000	0.0171	0.0290	0.0005	0.0284	0.0044	0.0001	0.0043

of output and this can be obtained directly from our SAM of the year 2006–2007. This is also called energy intensity with response to gross output of the sector. To estimate the indirect-induced effect, we have subtracted this direct effect from the total effects.

If we look at the impact on coal energy use for thermal electricity sector (NHY), we can see that the total effect on coal energy use is higher for the NHY sector (see Table 4.2). The total effect of NHY sector on coal is 0.2098 out of which 0.1131 is the direct and 0.0967 is the indirect-induced effect on coal energy use. This implies that if due to any exogenous changes the output of the NHY sector increases by one unit, the total coal energy requirement of the NHY sector will rise directly by the amount 0.1131 units and indirectly by the amount 0.0967 units.

Apart from the thermal electricity sector, the total impact on coal is observed higher for the iron and steel sector and the cement sector in comparison to other sectors of the economy (see Table 4.2). The total impacts on coal for these two sectors are 0.13 and 0.11, respectively, out of which the direct impacts are 0.07 and 0.06, respectively. Therefore, if the output of the iron and steel sector increases by one unit due to any exogenous changes, the economy wide demand for coal will be increased by 0.13 unit. The same logic is also valid for the cement sector.

By contrast, Table 4.2 shows that the direct and indirect-induced impact on coal is significantly low for agriculture and other services sector. The direct uses of coal in these sectors are negligible. Therefore, whatever increase in coal demand is observed for these sectors is due to indirect-induced effect.

In case of crude oil, we can observe from Table 4.2 that only the petroleum sector has significant direct impact on crude oil. This direct effect is negligible for some sectors and for most of the sectors this is almost zero. But when we take into account indirect-induced effect, the total impact on crude oil demand in the economy will increase significantly. For the thermal electricity sector and for most of the manufacturing sectors, we have observed significant impact on crude oil demand (see Table 4.2). We can say that though there are negligible direct impacts in some sectors but due to indirect-induced effect, their output growth will increase the economy wide demand for energy.

If we compare the direct and indirect-induced impact on different energy types, we can see this impact is higher for crude oil relative to other commercial energy type available from primary conventional sources (see Table 4.2). This impact is much lower in case of energy like hydro, natural gas, and nuclear. So, we can say that crude oil is the key source of primary energy in India.

However, the direct and indirect-induced impact on energy use and GHG emissions are observed due to the backward and forward linkage of the production sectors of the economy. The economic activity in a sector with high backward linkage provides stimulus to other sectors by requiring more inputs, whereas activity in a sector with high forward linkage stimulates higher outputs in other sectors by providing more inputs to them (Pradhan et al. 2006).

For illustration purpose, let us analyze the forward and backward linkage effect with an example of thermal electricity sector. The thermal electricity sector is a key source of the electric supply in case of Indian economy. The share of thermal electricity sector in total electric supply is about 86 % in the year 2006–2007.[1] So, the other production sectors of the economy are strongly dependent on the thermal electricity sector for their electricity requirement and hence the thermal electricity sector has a high forward linkage. Since there is high forward linkage, the expansion of thermal electricity sector will lead to expansion of the output of other sectors in the economy. Again the thermal electricity sector requires both energy and non-energy inputs for its production process and these are supplied by the other production sectors of the economy. If we look at the column of the thermal electricity sector from our environmentally extended SAM (ESAM), we can see that most of the nonenergy inputs of this sector are supplied from the energy-intensive sectors. Hence, total energy use in the energy-intensive sectors will increase due to expansion of thermal electricity sector.

Thus, we have demonstrated how the energy use will be increased if there is growth in sectoral output. Of course, this energy use causes GHG emissions in the atmosphere. According to Indian Network on Climate Change Assessment (INCCA) report, the energy-based GHG emissions is most in India. Almost 58 % of total carbon dioxide equivalent (CO_2EQ) emission is due to energy use (INCCA 2010). So our next objective is to analyze the impact on GHG emissions if the sectoral output grows due to any exogenous injection, and this is described in the following section.

4.5   Impact on GHG Emissions due to Sectoral Growth Resulting from any Exogenous Changes in the Economy

In this case, we have used the pollutant × activity block of pollution trade-off multiplier to analyze the impact of sectoral growth on the pollution emission, and this is given in Table 4.3. Each cell entry of Table 4.3 shows the direct and indirect-induced effect on generation of the pollutants due to change in the output of the production sectors. The higher value of multiplier in Table 4.3 implies higher impact on pollution generation. Therefore, the analysis would help us to find out the leading sectors of the economy, which have highest impact on the environment.

As Table 4.3 shows, the direct, indirect-induced impact on CO_2EQ emission is highest (i.e., 86.88 t) for thermal electricity sector. Out of this total impact, the direct impact of thermal electricity sector is about 49.22 t (Table 4.3). Therefore, if the output of the thermal electricity sector is increased by one unit, the total CO_2EQ generation in India will rise by 86.88 t. Hence, the thermal electricity sector of India is the leading sector in terms of GHG emissions and the same thing is also evident from INCCA report (INCCA 2010). INCCA report also says that the CO_2EQ emission from thermal electricity sector is high due to coal use in its production process.w The coal constitutes about 90 % of the total fuel mix used in the thermal electricity sector. So, the increase in output in thermal electricity sector will have

---

[1] Source: Our SAM.

Table 4.3 Impact of output growth on greenhouse gas (GHG) emissions (tons/lakhs of rupees of output). (Source: Author's estimation)

Sectors	CO_2 emission			CH_4 emission			N_2O emission			CO_2 EQ emission		
	Total effect	Direct effect	Indirect-induced effect	Total effect	Direct effect	Indirect-induced effect	Total effect	Direct effect	Indirect-induced effect	Total effect	Direct effect	Indirect-induced effect
PAD	18.677	0.775	17.902	0.515	0.222	0.293	0.006	0.003	0.003	31.270	6.285	24.984
WHT	19.608	0.839	18.769	0.205	0.000	0.205	0.007	0.003	0.003	26.071	1.923	24.148
CER	16.318	0.548	15.770	0.219	0.000	0.219	0.004	0.001	0.002	22.005	0.922	21.083
CAS	16.047	0.277	15.770	0.226	0.000	0.226	0.004	0.002	0.002	22.085	0.866	21.220
ANH	17.837	2.314	15.522	0.682	0.478	0.203	0.002	0.000	0.002	32.889	12.375	20.515
FRS	5.672	0.000	5.672	0.073	0.000	0.073	0.001	0.000	0.001	7.450	0.000	7.450
FSH	14.964	0.000	14.964	0.193	0.000	0.193	0.002	0.000	0.002	19.654	0.000	19.654
COL	13.355	0.000	13.355	0.281	0.128	0.153	0.002	0.000	0.002	19.780	2.691	17.088
OIL	3.480	0.000	3.480	0.046	0.004	0.042	0.000	0.000	0.000	4.581	0.078	4.503
GAS	9.323	0.000	9.323	0.510	0.397	0.113	0.001	0.000	0.001	20.409	8.337	12.072
FBV	15.677	0.743	14.934	0.210	0.000	0.210	0.003	0.000	0.003	20.885	0.749	20.136
TEX	15.989	0.066	15.923	0.187	0.001	0.187	0.002	0.000	0.002	20.585	0.085	20.500
WOD	14.778	0.296	14.482	0.175	0.000	0.175	0.002	0.000	0.002	19.044	0.298	18.747
MIN	5.225	0.124	5.101	0.061	0.000	0.061	0.001	0.000	0.001	6.699	0.124	6.575
PET	8.375	1.027	7.348	0.086	0.000	0.086	0.001	0.000	0.001	10.455	1.030	9.426
CHM	13.411	0.533	12.878	0.147	0.000	0.147	0.002	0.000	0.001	17.136	0.642	16.495
PAP	14.496	1.306	13.190	0.147	0.006	0.141	0.002	0.000	0.002	18.073	1.446	16.628
FER	18.142	4.846	13.296	0.169	0.000	0.169	0.002	0.000	0.002	22.240	4.885	17.355
CEM	50.249	34.669	15.579	0.151	0.000	0.151	0.002	0.000	0.002	53.887	34.669	19.218
IRS	18.800	3.995	14.805	0.156	0.001	0.156	0.002	0.000	0.002	22.564	4.018	18.546
ALU	6.596	0.209	6.387	0.071	0.000	0.071	0.001	0.000	0.001	8.299	0.209	8.090
OMN	9.525	0.545	8.980	0.097	0.000	0.097	0.001	0.000	0.001	11.893	0.548	11.344
MCH	12.181	0.263	11.918	0.128	0.000	0.128	0.001	0.000	0.001	15.309	0.264	15.044
NHY	80.212	47.827	32.385	0.278	0.065	0.213	0.003	0.001	0.002	86.883	49.422	37.461
HYD	15.383	0.000	15.383	0.198	0.000	0.198	0.002	0.000	0.002	20.198	0.000	20.198
NUC	24.600	0.000	24.600	0.200	0.000	0.200	0.002	0.000	0.002	29.446	0.000	29.446

Table 4.3 (continued)

Sectors	CO_2 emission			CH_4 emission			N_2O emission			CO_2 EQ emission		
	Total effect	Direct effect	Indirect-induced effect	Total effect	Direct effect	Indirect-induced effect	Total effect	Direct effect	Indirect-induced effect	Total effect	Direct effect	Indirect-induced effect
BIO	15.452	0.000	15.452	0.201	0.000	0.201	0.002	0.000	0.002	20.355	0.000	20.355
WAT	17.019	0.000	17.019	0.197	0.000	0.197	0.002	0.000	0.002	21.816	0.000	21.816
CON	16.794	0.001	16.793	0.173	0.000	0.173	0.002	0.000	0.002	21.006	0.001	21.005
LTR	15.703	2.366	13.337	0.162	0.000	0.161	0.002	0.000	0.002	19.689	2.412	17.277
RLY	20.731	0.698	20.033	0.185	0.000	0.185	0.002	0.000	0.002	25.324	0.782	24.541
AIR	23.815	9.609	14.207	0.171	0.000	0.171	0.002	0.000	0.002	28.091	9.693	18.398
SEA	15.745	0.830	14.916	0.184	0.000	0.184	0.002	0.000	0.002	20.257	0.839	19.418
HLM	15.045	0.000	15.045	0.189	0.000	0.189	0.002	0.000	0.002	19.677	0.000	19.677
SER	14.972	0.007	14.964	0.193	0.000	0.193	0.002	0.000	0.002	19.670	0.007	19.663

significant direct effect on total CO_2EQ emission in India. In the above section, we have discussed about the backward and forward linkage of thermal electricity sector. Here also, we have seen that backward and forward linkage affect significantly GHG emissions for thermal electricity sector.

After thermal electricity sector, the total impact on CO_2EQ emission is observed high for cement sector. As Table 4.3 shows, if the output of the cement sector increases by ₹ 1 lakh, the total CO_2EQ emission in India will increase by 53.88 t. The energy-based emission from the cement sector is almost 44%, whereas production-process-based CO_2EQ emission is 56% (INCCA 2010). Therefore, an increase in output of the cement sector will increase directly due to both energy use and production process. On the other hand, we have observed significant impact on coal use due to backward and forward linkage effect (see Table 4.2). Therefore, we have observed for the cement sector significant direct and indirect-induced impact on CO_2EQ emission.

On the other hand, the values of the direct pollution coefficients are very small for the agricultural sectors and other services sectors of the Indian economy (Table 4.3). If we look at the direct effect of paddy sector, we find that the CO_2EQ emission per unit of paddy output is about 6.29 t per ₹ 1 lakh of output. However, it is observed from Table 4.3 that the total effect on CO_2EQ emission generation is significant (i.e., 32.17) in paddy sector. In case of service sector, Table 4.3 indicates that the direct effect is negligible. But due to indirect-induced effect, the CO_2EQ emission in India will increase by 19.67 t resulting from an increase in service sector output by ₹ 1 lakh. Hence, we can conclude that those sectors that have small direct impact may have significant indirect and induced effect on the GHG emissions.

The above discussion indicates that the energy-intensive sector has the significant impact on energy use and GHG emissions. But it will be wrong to say that the growth in non energy-intensive sector will not have significant impact on energy use as well as on GHG emissions. This could happens because there are backward and forward linkage effects between the sector and the rest of the economy due to which there coould be significant impact on energy use and GHG emissions.

Till now, we have described the impact of output growth on GHG emissions in India. But this growth in output will increase the sectorwise employment in the economy. In the context of economic growth analysis in India, the analysis of employment change is essential. On the other hand, it will be interesting to see whether this increase in employment have adverse impact on GHG emissions in India or not. This issue is addressed in the following section.

4.6   Impact of Change in Employment on GHG Emissions

To analyze the impact of sector-specific employment change on GHG emissions, we have used method of employment multiplier. Before we estimate this employment multiplier, let us see the pattern of sectorwise employment for the year 2006–2007, and this is given in following Table 4.4.

Table 4.4 Sectorwise share of employment and labor intensity (2006–2007). (Source: www.india-stat.com and author's estimation)

Activity sector	Share of employment	Labor intensity (number of persons per lakh of rupees of output)
Agriculture	0.685	2.623
Energy	0.018	0.088
Manufacturing	0.114	0.105
Transport	0.039	0.243
Other services	0.145	0.245
Total	1.000	

As this table shows, the share in total employment and labor intensity is high in the agriculture sector and other services sector. It is interesting to note that the share of employment as well as labor intensity is higher in other services sector in comparison to the manufacturing sector. Therefore, if sectoral output changes due to unitary changes in the exogenous account of SAM, the employment opportunity will increase more in agriculture and other services sector relative to manufacturing and other sectors of the economy. If we link GHG emissions with this employment growth, we can find the sector for which the impact of increase in GHG emissions is less as compared to other sector. To do this, we have first estimated sectorwise employment multiplier and then we relate that with the GHG emissions. The methodology and analysis is described below.

4.6.1   Employment Multiplier

Employment multiplier shows the direct and indirect changes in employment if the output of a sector changes due to any exogenous changes in the economy. Below, we have estimated the employment multiplier to show the direct and indirect-induced impact on employment due to any exogenous changes in the economy.

Let L_i be the employment in sector i and E_i be fixed employment coefficient. Therefore,

$$E_i = L_i / Y_i \qquad (4.12)$$

Reorganizing this equation with Eq. 4.6 by substituting MX for Y, we may rewrite it as

$$L = eMX \qquad (4.13)$$

Where e is the diagonal matrix formed with elements E_i and the element of "e" corresponding to nonproduction account are zero.

Differentiating the above equation we get,

$$\partial L / \partial X = eM = R \qquad (4.14)$$

Therefore, R_i gives us the employment multiplier for sector i which indicates the direct, indirect, and induced employment created in the whole economy when the exogenous variable changes by one unit.

4.6.2   Impact of Employment Creation on GHG Emissions

To investigate the impact of employment creation on GHG emissions, we have to link this employment multiplier with the GHG emissions. To link GHG emissions with employment multiplier, we have followed the same method as we have applied for pollution trade-off multiplier. Here, we have estimated sectorwise emission employment ratio by dividing sectorwise total emission from their corresponding level of employment (i.e., number of labor engaged in that particular sector). Premultiplying this with the employment multiplier, we have obtained the employment-emission multiplier and this is shown in the following equation.

$$E_p = e_p \cdot R \qquad (4.15)$$

where e_p is the employment emission ratio.

Thus, we have estimated the impact on sectoral GHG emissions due to increase in sectoral employment resulting from unitary changes in the exogenous account in our SAM. The result obtained in this way is given in Table 4.5.

From the above table, we can see that the value of employment multiplier is significant for agriculture and service sector relative to other sectors of the economy. This is due to their labor intensity as shown in Table 4.4. If we look at the impact on GHG emissions, we find that the increase in GHG emissions is not significant for service sector in comparison to the manufacturing and other sectors of the economy. In case of manufacturing sector, though its value of employment multiplier is less, it has significant impact on GHG emissions.

Thus, we have seen that the sectorwise employment will change at a different rate due to unitary changes in exogenous variable for all sectors. Also, we have seen that the impacts of employment change on GHG emissions are different for different sectors. This difference in impact is due to the sectorwise difference in energy consumption. In this case, it will be interesting to see the difference in energy consumption per unit of labor in the production sector. Here it is assumed that the energy consumption per unit of labor for the group of sectors will be followed by their counterpart. In Table 4.6, we have estimated energy consumption per unit of labor for broad group of sector in India for the year 2006–2007.

It is clear from Table 4.6 that the payment for energy use per unit of labor is highest for energy production sector itself. In case of manufacturing sector, the payment for electricity use per unit of labor is almost 13 times higher than the other services sector. The payment for thermal electricity consumption is almost negli-

Table 4.5 Impact of employment change on GHG emissions. (Source: Author's estimation)

Production sector	Employment multiplier	CO_2 (t)	CH_4 (t)	N_2O ('t)	CO_2EQ (t)
PAD	3.718	1.466	0.421	0.005	11.897
WHT	4.225	2.002	0.000	0.008	4.587
CER	5.601	1.024	0.000	0.002	1.723
CAS	4.637	0.435	0.000	0.003	1.361
ANH	4.031	3.669	0.758	0.000	19.618
FRS	2.745	0.000	0.000	0.000	0.000
FSH	4.222	0.000	0.000	0.000	0.000
COL	0.912	0.000	0.331	0.000	6.948
OIL	1.824	0.000	0.071	0.000	1.497
GAS	1.179	0.000	0.944	0.000	19.822
FBV	0.603	2.150	0.001	0.000	2.167
TEX	0.694	0.222	0.003	0.000	0.287
WOD	0.723	0.467	0.000	0.000	0.470
MIN	1.125	0.964	0.000	0.000	0.967
PET	1.803	328.319	0.035	0.001	329.273
CHM	2.032	16.023	0.008	0.010	19.278
PAP	0.288	6.343	0.031	0.000	7.020
FER	1.821	40.496	0.004	0.001	40.819
CEM	0.138	105.931	0.000	0.000	105.931
IRS	0.439	64.020	0.008	0.001	64.382
ALU	0.257	10.158	0.000	0.000	10.159
OMN	1.144	15.463	0.000	0.000	15.563
MCH	0.856	4.440	0.000	0.000	4.469
NHY	0.281	965.130	1.321	0.014	997.319
HYD	1.291	0.000	0.000	0.000	0.000
NUC	0.178	0.000	0.000	0.000	0.000
BIO	4.184	0.000	0.000	0.000	0.000
WAT	0.661	0.000	0.000	0.000	0.000
CON	1.039	0.006	0.000	0.000	0.006
LTR	1.700	16.978	0.003	0.001	17.306
RLY	0.712	2.020	0.000	0.001	2.263
AIR	0.524	9.679	0.000	0.000	9.764
SEA	0.414	1.330	0.000	0.000	1.344
HLM	0.437	0.000	0.000	0.000	0.000
SER	4.523	0.132	0.000	0.000	0.133

Table 4.6 Energy cost per unit of labor employed (energy in rupee value/labor). (Source: Author's estimation with the help of SAM of the year 2006–2007)

Production sector	Rupees of primary energy use/labor	Rupees of petroleum energy use/labor	Rupees of thermal electricity use/labor
Agriculture	1.36	544.01	298.17
Energy	306,828.51	41,186.32	68,904.85
Manufacturing	11,922.10	19,506.07	13,074.87
Transport	27.33	82,546.33	4787.66
Services	71.87	1796.37	1164.83

Table 4.7 Sector-specific share in gross domestic product at factor cost. (Source: Author's estimation with the help of national account statistics 2009)

Year	Agriculture	Mining	Manufacturing	Electricity	Construction	Transport	Other service
1999–2000	24.99	2.33	14.78	2.49	5.71	7.47	42.23
2000–2001	23.89	2.28	15.26	2.44	5.81	7.96	42.35
2001–2002	23.99	2.20	14.79	2.34	5.71	8.15	42.81
2002–2003	21.43	2.30	15.22	2.36	5.94	8.96	43.79
2003–2004	21.72	2.19	14.95	2.28	6.13	9.52	43.21
2004–2005	20.20	2.20	15.12	2.29	6.62	10.25	43.32
2005–2006	19.56	2.11	15.06	2.19	7.05	10.74	43.29
2006–2007	18.51	2.04	15.39	2.12	7.20	11.42	43.32

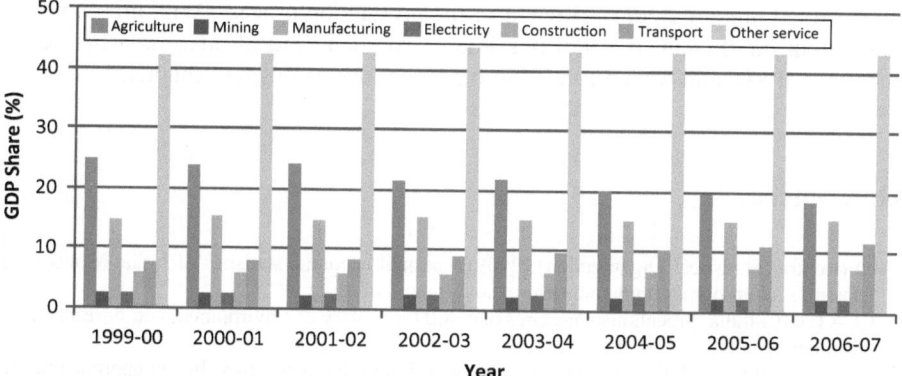

Fig. 4.1 Sector-specific shares in gross domestic product. (Source: CSO (2009), National Accounts Statistics)

gible for agriculture sector. Again in case of other forms of energy, the payment for energy use per unit of labor is significantly less than the manufacturing sector. As a result, despite the low employment multiplier the impact on GHG emissions is high for manufacturing sector as compared to the other services sector in India.

Furthermore, it can be observed from Table 4.7 and Fig. 4.1 that the share of other services sector in gross domestic product (GDP) is highest in India. In the year 2006–2007, the share of service sector is about 43 %, whereas the same for agriculture sector is 18 % and manufacturing sector is 15.59 %. On the other hand, the share of service sector is gradually increasing over the years, whereas this is declining for agricultural sector. In case of manufacturing, the share in GDP is almost constant during the period 1999–2000 to 2006–2007. Therefore, it is evident from Table 4.7 that Indian economy is structurally biased towards service sector. Hence, we can say that if the Indian economy follows this structure in future, the growth in service sector will boost economic growth, create significant employment opportunities with less impact on GHG emissions.

Hence, this study shows that the impact on GHG emissions of the economy is not only dependent on the sector-specific energy intensity but also due to the backward and forward linkages between the activities. As a consequence of the linkage effect, the growth in output of low-energy-intensive sectors or low-emission-intensive sectors may have significant indirect-induced impact on total GHG emissions. On the other hand, if the objective is to create green employment in India, service sector growth will be the key as it has higher labor intensity with less GHG emissions intensity.

However, if backward and forward linkages between the sectors change, the indirect-induced impact may change, resulting in differential impact on GHG emissions. On the other hand, if any technological improvement changes the emission intensity, then direct impact on the GHG emissions may change. Now, this change in backward and forward linkage between the sectors will change if there is change in input–output coefficients. But the SAM multiplier model in this chapter is based on the fixed input–output coefficient of the production process. With this fixed input–output coefficient, it is not possible to do such kind of analysis. This needs further analysis, and we have taken into account this in the next chapter.

References

CSO (Central Statistical Organization) (2009) National accounts statistics of India. Ministry of statistics and program implementation, Government of India

INCCA (2010) India: greenhouse gas emission 2007. Ministry of Environment and Forests, Government of India

Planning Commission (2011) Faster, sustainable and more inclusive growth—an approach to the 12th five year plan, Government of India

Pradhan BK, Saluja MR, Singh SK (2006) A social accounting matrix for India, concepts, construction and applications. Sage, New Delhi

Robert K (1975) Input output analysis and air pollution control. In: Edwin SM (ed) Economic analysis of environmental problems. NBER, Cambridge, pp 259–274

UNEP (2008) Green jobs: towards decent works in a sustainable, low carbon world. United National Environment Programme. United Nations Office, Nairobi. http://www.unep.org/civilsociety/Implementation/GreenJobs/tabid/104810/Default.aspx

Chapter 5
Greenhouse Gas (GHG) Emissions in India—A Structural Decomposition Analysis

This chapter describes the historical trend of greenhouse gas (GHG) emissions and analyzes the factors influencing this historical trend. Since every economy passes through various structural change processes over time, their implications must be assessed for future policy implementation. Therefore, in this chapter, we have described the methodology of input-output (IO) structure decomposition analysis to assess and determine the factors for historical GHG emissions trend in the Indian economy.

5.1 Concept of Structural Decomposition Analysis

The structure and level of both production and consumer demand of a country determine the amount of GHG emissions released back into the atmosphere. In the last two decades, reduction in carbon dioxide equivalent (CO_2EQ) emission from economic activities for decoupling economic growth and GHG flow has been incorporated into the climate mitigation policy agenda globally (Dell et al. 2008). To define appropriate policy intervention, understanding of how emissions are generated and the economic and technological factors drive and delineate the country's GHG profile is pertinent. This necessitates in-depth study in this particular issue. Therefore, in this chapter, our objective is to analyze the driving factors in India's GHG emissions by decomposing overall emissions into four key determinants: intensity effect, effect of technical coefficient change, final demand mix effect, and final demand level effect over a time gap of 12 years, from 1994–1995 to 2006–2007. By intensity effect, we mean change in emission intensity during this time period. The effect of technical coefficient change implies the change in intermediate use pattern or change in IO coefficients in the production process during this time period. The final demand mix change is nothing but change in sector-wise share in total final demand of the economy over the time gap of 12 years. Finally, the final demand-level effects describe the effect of real change in aggregate level of final demand in the economy during the year 1994–1995 to 2006–2007.

B. D. Pal et al., *GHG Emissions and Economic Growth*,
India Studies in Business and Economics, DOI 10.1007/978-81-322-1943-9_5,
© Springer India 2015

However, in the SAM multiplier analysis of Chap. 4, the household final demand is endogenously determined due to any exogenous change into the economy. But in this chapter, we consider this as the exogenous driving force into the economy. Therefore, to do this decomposition analysis, we have applied the IO structural decomposition analysis (SDA) to the emission data constructed in previous chapters through environmentally extended social accounting matrix (ESAM) for India for the year 2006–2007.

5.2   Decomposition Method

Recently, a number of studies have concentrated on energy and environment (GHG emissions) by applying IO SDA. Work in this area has been done by Lin and Chang (1996). They used Divisia Index to classify emission changes of SO_2, NOx, and CO_2 from major economic sector in Taiwan during 1980–1992. Very few works have been done in India especially using IO method. Roy (2007) has done decomposition analysis using Laspeyers method. Gupta and Hall (1997) estimated GHG emissions for 3 reference years 1980–1981, 1985–1986, and 1987–1988 using simple spreadsheet model. In the context of Indian economy using IO data, the latest study is done by Mukhopadhyay (2001). This study applied IO structural decomposition analysis to decompose total change in CO_2 emission into nine factors—real added value, the level of CO_2 intensity, CO_2 substitution to energy mix, the rate of domestic production to intermediate input, the mix of intermediate input, the level of domestic final demand, and domestic final demand mix. This study estimates CO_2 emission change for the years 1973–1974, 1983–1984, 1991–1992, and 1996–1997. In this study, we have followed this method with the help of monetary IO table for India to analyze the change in GHG emissions between the year 1994–1995 and 2006–2007. The method is given below.

Mathematically, the structure of IO model can be expressed as,

$$Y = AY + sX \tag{5.1}$$

The solution of Eq. 5.1 gives

$$Y = (I - A)^{-1} sX \tag{5.2}$$

where,

Y Gross output
X Volume of aggregate final demand
s Consumption structure
I Identity matrix

and

$(I–A)$ The matrix of total input requirements.

The above IO model is used to compute the amount of CO_2 emission that takes place in the production of various activity levels. The amount of CO_2 can be obtained in the following way.

$$E = pY \tag{5.3}$$

where, p is the emission intensity, i.e., emission per unit of output and E is the level of GHG emissions.

Next, the model develops a structural decomposition analysis. It is an ideal technique to study changes over a period of time. It has become a major tool for disentangling the growth in some variables over time, separating the changes in the variable into its constituent parts. SDA seeks to distinguish major tool for disentangling the growth in some variables over time, separating the changes in the variable into its constituent parts. SDA seeks to distinguish major sources of change in the structure of the economy broadly defined by means of a set of comparative static changes in key parameters of an IO table.

So, the change in total GHG (CO_2 plus CO_2EQ N_2O and CH_4) emission between any 2 years, i.e., initial year (denoted by "0") and the terminal year (denoted by "1") can be identified as:

$$\Delta E = E1 - E0 \tag{5.4}$$

Substituting Eq. 5.3 in Eq. 5.4,

$$\Delta E = p1Y1 - p0Y0 \tag{5.5}$$

Substituting Eq. 5.1 in Eq. 5.5,

$$
\begin{aligned}
\Delta E = {} & p_1 (I-A)^{-1}{}_1 s_1 X_1 + p_1 (I-A)^{-1}{}_0 s_0 X_0 - p_0 (I-A)^{-1}{}_0 s_0 X_0 - p_1 (I-A)^{-1}{}_0 s_0 X_0 \\
& + p_0 (I-A)^{-1}{}_1 s_0 X_0 - p_0 (I-A)^{-1}{}_0 s_0 X_0 - p_0 (I-A)^{-1}{}_0 s_0 X_0 + p_0 (I-A)^{-1}{}_0 s_1 X_0 \\
& - p_0 (I-A)^{-1}{}_0 s_0 X_0 - p_0 (I-A)^{-1}{}_0 s_1 X_0 + p_0 (I-A)^{-1}{}_0 s_0 X_1 - p_0 (I-A)^{-1}{}_0 s_0 X_0 \\
& - p_0 (I-A)^{-1}{}_0 s_0 X_1 + p_0 (I-A)^{-1}{}_0 s_0 X_0
\end{aligned}
\tag{5.6}
$$

$$
\begin{aligned}
\Delta E = {} & p_1 (I-A)^{-1}{}_0 s_0 X_0 - p_0 (I-A)^{-1}{}_0 s_0 X_0 + p_0 (I-A)^{-1}{}_1 s_0 X_0 - p_0 (I-A)^{-1}{}_0 s_0 X_0 \\
& + p_0 (I-A)^{-1}{}_0 s_1 X_0 - p_0 (I-A)^{-1}{}_0 s_0 X_0 + p_0 (I-A)^{-1}{}_0 s_0 X_1 - p_0 (I-A)^{-1}{}_0 s_0 X_0 \\
& + p_1 (I-A)^{-1}{}_1 s_1 X_1 - p_1 (I-A)^{-1}{}_0 s_0 x_0 - p_0 (I-A)^{-1}{}_1 s_0 X_0 - p_0 (I-A)^{-1}{}_0 s_1 X_0 \\
& + p_0 (I-A)^{-1}{}_0 s_0 X_0
\end{aligned}
\tag{5.7}
$$

Separating Eq. 5.7 into five parts,

$$\Delta E = p_1 (I-A)^{-1}_0 s_0 X_0 - p_0 (I-A)^{-1}_0 s_0 X_0 \qquad (5.7a)$$

$$+ p_0 (I-A)^{-1}_1 s_0 X_0 - p_0 (I-A)^{-1}_0 s_0 X_0 \qquad (5.7b)$$

$$+ p_0 (I-A)^{-1}_0 s_1 X_0 - p_0 (I-A)^{-1}_0 s_0 X_0 \qquad (5.7c)$$

$$+ p_0 (I-A)^{-1}_0 s_0 X_1 - p_0 (I-A)^{-1}_0 s_0 X_0 \qquad (5.7d)$$

$$+ \left(\begin{array}{l} p_1 (I-A)^{-1}_1 s_1 X_1 - p_1 (I-A)^{-1}_1 s_0 x_0 - p_0 (I-A)^{-1}_1 s_0 X_0 \\ -p_0 (I-A)^{-1}_0 s_1 X_0 + p_0 (I-A)^{-1}_0 s_0 X_0 \end{array} \right) \qquad (5.7e)$$

Mathematical expression (5.7a) shows the CO_2 emission changes due to changes of CO_2 intensity. Expression (5.7b) reflects the CO_2 emission changes due to the changes in technical coefficient matrix. Expression (5.7c) refers CO_2 emission changes due to changes in share of final demand of various sectors. Expression (5.7d) defines CO_2 emission changes due to changes in the volume of demand. Finally expression (5.7e) represents the total joint effects when two or more of the above factors occurred simultaneously.

5.2.1 Data and Empirical Estimation

To conduct the structural decomposition analyses, following data are required:

a. The IO table with same sectoral classification for the year 1994–1995 to 2006–2007. The purpose of this two IO table is to solve Eq. 5.7.
b. The data on prices indices are required to convert the volume of final demand at current prices into constant prices. As our analysis is restricted for 1994–1995 and 2006–2007, therefore, we need prices indices for these 2 years with same base year.
c. Data on sector-wise CO_2 emission between 1994–1995 and 2006–2007 are required for this analysis.

We have 35-sector IO table for the year 2006–2007 used at the time of constructing 35 sectors SAM for India for the year 2006–2007. So we have to get the IO table for the year 1994–1995 with same sectoral classification of the year 2006–2007. In this study, we have obtained a 60-sector IO table for Indian from the study by Pradhan et al. (1999). This IO table is based on the IO table published by the Central Statistical Office (CSO). But the sectoral classification of this IO table does not matched with the 35 sectors of our SAM. So, we have to aggregate the 60-sector IO table to get the 35-sector IO table for India for the year 1994–1995. The method of aggregation is described below.

5.2.2 Aggregation of 60-Sector IO Table of the Year 1994–95

The steps involved in aggregating the 60 sectors IO table of the year 1994–1995 are as follows:

Step 1 In the first step, we have made a map of concordance between the 35 sectors of the 2006–2007 IO table and the 60 sectors of the 1994–1995 IO table. The map of concordance is given in the following Table 5.1.

As Table 5.1 shows, there are some sectors in the 2006–2007 IO table which are partly matched with the sectors of 1994–1995 IO table. Also there are some new sectors in the 2006–2007 IO table. So to obtain the 35-sectors IO table of the year 1994–1995, we have to construct separate rows and columns of these sectors for the year 1994–1995. Now the methods are described in the following steps.

Step 2 In this step, we have first obtained the separate rows and columns for the sectors which are partly matched (except the new sectors introduced in the 2006–2007 IO table) with the sectors of 1994–1995 IO table. Now to do this, we have referred the 115-sectors IO table of the year 1993–1994 as published by CSO. Thus, we have obtained a 30-sectors IO table of the year 1994–1995.

Step 3 Now in the third step, we have extended the 30-sectors IO table of the year 1994–1995 by extending the electricity and transport sector. In this case, we have used the relevant data from the following sources:

a. To obtain separate row and column of the hydro electricity sector, non-hydro electricity sector, and the nuclear electricity sector, the data for the year 1994-95 are obtained from the annual reports of the Hydro Electric Corporation of India (www.nhpcindia.com), the Thermal electricity corporation of India (www.ntpc.co.in) and the Nuclear power Corporation of India (www.npcl.co.in).
b. The row and column of the sea transport sector and air sector are obtained on the basis of data available from the annual reports of different official transport authorities of India.

Step 4 In the last step, we have constructed the row and column of the biomass sector. Now to construct the row and column of the biomass sector, we have followed the same method as we have followed in case of 2006–2007 SAM (see Chap. 3). But in this case, we did not get data on biomass originated from industries. So we have used share of paper and food and beverages industries in biomass production of the year 2006–2007 to get the biomass output of the year 1994–1995. But in case of firewood and agricultural residual the data are obtained from the National Accounts Statistics (NAS) data. Hence in this way, we have obtained the 35-sector IO table for the year 1994–1995 and this is given in Appendix 6.

Next, we have to convert the aggregate volume of final demand at current prices into constant prices. To do this, data are obtained from NAS published by CSO.

Finally, we have to estimate the sector-wise emission coefficients for the year 1994–1995 for the sector described in 35-sector IO table. In this case, we have ob-

Table 5.1 Mapping between 2006–2007 IO sectors and the 1994–1995 IO sectors

		2006–2007 IO sectors	1994–1995 IO sectors
1	PAD	Paddy rice	Part of S1
2	WHT	Wheat	Part of S1
3	CER	Cereal, grains, etc, other crops	Part of S1 and S4
4	CAS	Cash crops	S2,S3
5	ANH	Animal husbandry and production	S5
6	FRS	Forestry	S6
7	FSH	Fishing	S7
8	COL	Coal	S8
9	OIL	Oil	Part of S9
10	GAS	Gas	Part of S9, part of S47
11	FBV	Food and beverage	S12, S13, S14, S15
12	TEX	Textile and leather	S16, S17, S18, S19, S24
13	WOD	Wood	S20
14	MIN	Minerals n.e.c.	S10, S11
15	PET	Petroleum and coal production	S26, S27
16	CHM	Chemical, rubber, and plastic	S23, S25, S28, S29, S31, S32
17	PAP	Paper and paper production	S22
18	FER	Fertilizers and pesticides	S30, Part of S32
19	CEM	Cement	S33
20	IRS	Iron and steel	S35
21	ALU	Aluminum	S36
22	OMN	Other manufacturing	S21, S34, S37, S43, S44
23	MCH	Machinery	S38, S39, S40, S41, S42
24	HYD	Hydro	Part of S46
25	NHY	Thermal	Part of S46
26	NUC	Nuclear	Part of S46
27	BIO	Biomass	
28	WAT	Water	Part of S47
29	CON	Construction	S45
30	LTR	Road transport	Part of S49
31	RLY	Rail transport	S48
32	AIR	Air transport	Part of S49
33	SEA	Sea transport	Part of S49
34	HLM	Health and medical	S58
35	SER	All other services	S50, S51, S52, S53, S54, S55, S56, S57, S59, S60

Note: The description of 60 sectors is given in Appendix 5

tained the data on sector-wise GHG emissions from the National Communications Corporation Limited (NATCOM)-I report. The sectors given in this report are same as it is given in Indian Network on Climate Change Assessment (INCCA) report of the year 2010. Therefore, we have followed the same method as it was adopted in estimating sector-wise emission for the year 2006–2007 for ESAM (see Chap. 3). Thus, we have obtained required data for the decomposition analysis purpose. The data on GHG emissions obtained in the above-mentioned way can be used to see the observed changes in GHG emissions in India between the year 1994–1995 and 2006–2007 and this is shown in the following section.

Table 5.2 Observed changes in GHG emissions (thousand tons)

Broad sector of economy	1994–1995	2006–2007	Absolute change	Average annual change
Agriculture	344,482	334,411	−10,071	−0.25%
Mining	33,923	33,162	−761	−0.19%
Thermal electricity	355,033	739,704	384,671	4.33%
Manufacturing	322,892	445,726	122,834	2.30%
Transport	80,280	142,041	61,760	3.62%
Other services	15,958	1,583	−14,375	−75.66%
All Sector	1,152,569	1,696,628	544,059	2.67%

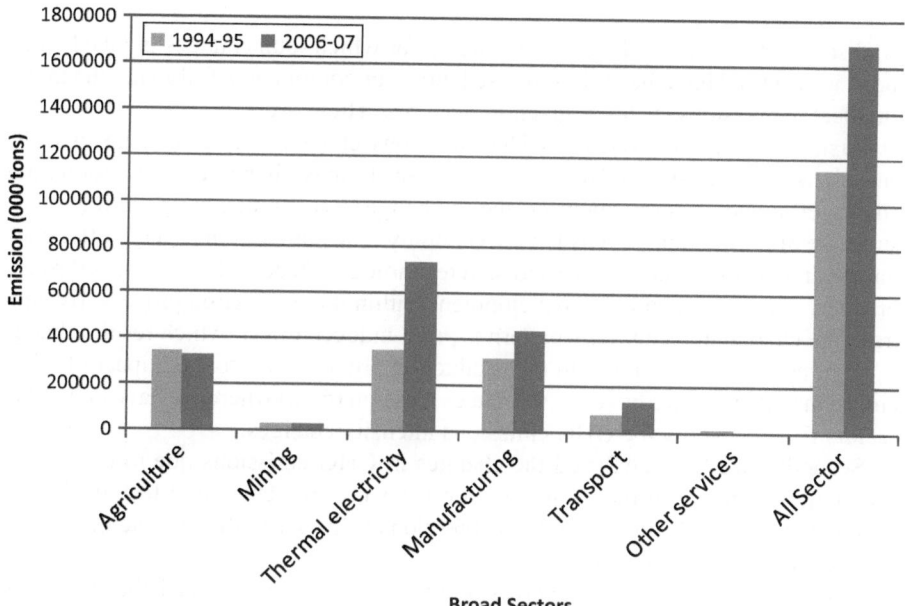

Fig. 5.1 Observed changes in GHG emissions. (Source: Authors' Estimate)

5.3   Observed Changes in GHG Emissions in India

Change in GHG emissions observed between the years 1994–1995 and 2006–2007 is shown in following Table 5.2 and Fig. 5.1.

It is observed from Table 5.2 and Fig. 5.1 that the total GHG emissions in India has increased by almost 544 million tons during this 12-years gap. The rising trend of GHG emissions are observed from the manufacturing, transport, and thermal electricity sector. For thermal electricity sector, it has increased at an annual rate of 4.33% and for transport sector average annual increase in GHG emissions is 3.60%. The average annual increase in GHG emissions for manufacturing sector is slightly lower (i.e., 2.30) than these two sectors. Between the year 1994–1995 and 2006–2007, the GHG emissions from agriculture sector has declined at an average rate of (−) 0.25 % annually. Again in case of service sector, the fall in GHG emissions is the highest (−) 75 %).

Now, let us see what are the factors for which we have this change in GHG emissions in India between the year 1994–1995 and 2006–2007, and this is described in the following section.

5.4   Findings from Structural Decomposition Analysis

To discuss empirical results of SDA, factors for which total changes in GHG emissions is observed have been decomposed into four components following the mathematical expressions (5.7a–d) given in Sect. 5.2. These are:

First, we have analyzed the GHG emissions changes due to change in GHG emissions intensity. This GHG emissions intensity may change due to autonomous energy efficiency improvement or due to change in technological pattern (i.e., IO structure of production activity) in the economy. The autonomous energy efficiency improvement is not due to price-induced technology effect; rather it occurs through practice, experience, and skill development within the production process. By contrast, the change in technology pattern is price-induced effect, which results change in input consumption pattern in the production process. To see the impact of GHG emissions intensity, we have considered expression (5.7a) where we have kept other factors fixed and only the GHG emissions intensity changes.

Secondly, we have analyzed the changes in GHG emissions due to changes in production structure in the economy, in other word, the change in technical coefficient matrix. This is done from the expression (5.7b). Here all other factors remain fixed only the $(I–A)^{-1}$ matrices changes.

Thirdly, with the help of expression (5.7c), we have analyzed the impact of change in consumption structure on GHG emissions change. In this case, only the sector-wise share in final demand has been changed but there is no change in the other factors. This, in other way, shows the effect of commodity substitution/change in households' behavior over the years. Fourthly, we have analyzed the impact of change in aggregate volume of final demand on GHG emissions change with the help of expression (5.7d). In this case, the aggregate level of final demand for the year 1994–1995 has been changed to 2006–2007 level. Here, we have taken these two final demands at constant 1994–1995 prices.

Now, the point to be noted here is that the volume of final consumption expenditure, in the national income accounting system, considers as proxy of total households income. So, considering volume of consumption expenditure at constant price for 2 different years shows the change in real households' income in the economy. Hence, the change in GHG emissions due to change in consumption volume can be treated as income effect on GHG emissions. On the other hand, the unit-free consumption share, when it changes over time, shows the commodity substitution

Table 5.3 Factor for which GHG emissions changes in India (unit thousand tons). (Source: Authors'Estimate)

	Emission intensity change	Technical coefficient change	Consumption structure change	Volume of consumption	Mixed effect	Total effect (GHG emissions of 2006–2007)	Observed change over 1994–1995 level
Agriculture	−19,1902	2,910	−117,730	634,663	6,471	334,412	−10,071
Mining	−18,163	11,728	−5,832	85,468	−40,040	33,161	−761
Thermal electricity	90,373	644,287	−66,857	3,392,022	−33,20122	739,703	384,671
Manufacturing	−13,8942	23,839	321	1,331,282	−77,0773	445,727	122,834
Transport	−5,670	15,727	31,896	396,477	−29,6389	142,041	61,760
Other services	−15,153	831	2,985	64,552	−51,632	1,583	−14,375
All sector	−279,456	699,322	−155,217	5,904,464	−4,472,484	1,696,629	544,059

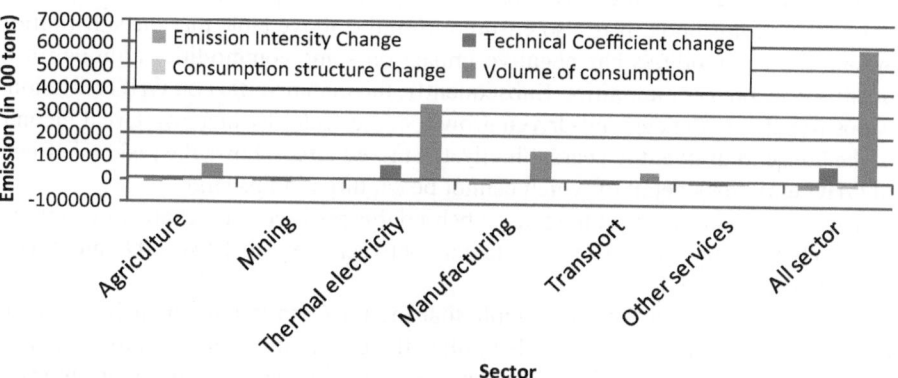

Fig. 5.2 Factor-wise GHG emissions change. (Source: Authors'Estimate)

effect on GHG emissions. Thus, this analysis helps us to understand the price effect through both the channel of substitution and income effect.

Now the sector-specific result of structural decomposition analysis is given in the following Table 5.3 and Fig. 5.2. In Table 5.2, we have shown this result for the broad group of sectors in India.

It is observed from the above Table 5.3 that the change in emission intensity has negative impact on GHG emissions for almost every sector of the economy. This implies that there is an improvement in autonomous energy efficiency over the years. However, the positive impact of energy intensity corresponding to thermal electricity sector ensures that energy efficiency has not been improved much in this sector over the years, causing increases in GHG emissions inventory. The impact of change in technical coefficients is positive for all the sectors. This happens due to interdependency/linkages among the sectors. Due to these linkages, even if one

Table 5.4 Change in emission intensity, technical coefficients, and consumption share between 1994–1995 and 2006–2007

	Primary energy	Thermal electricity	Other commercial energy	Nonenergy input	Final consumption share	Real change in GHG intensity (tons/Rs. lakh of output)
Agriculture	−0.0008	0.0032	0.0084	0.0262	−0.1177	−0.0054
Mining	−0.0007	−0.0180	−0.0200	−0.0805	−0.0314	−0.0078
Thermal electricity	0.5476	−0.1566	−0.0646	−0.5299	−0.0037	0.0212
other energy	0.0767	0.0348	0.0457	−0.4577	0.0127	−0.0229
Manufacturing	0.0044	−0.0109	0.0044	−0.0203	0.0019	−0.0004
Transport	−0.0005	−0.0013	0.0724	−0.0926	0.0274	−0.0005
Services	−0.0026	−0.0081	−0.0007	−0.0732	0.0786	−0.0003

sector improves technology led energy efficiency, its linkages with other sectors (which may be energy intensive) ensures the positive impact as we have explained in Chap. 4 of this book. It is also ensured from the above table that the substitution among the commodities has negative impact on some commodities although it is positive for some commodities. Subsequently, the income effect on GHG emissions is positive due to the demand-driven growth in the sector by increasing real income of the households over the years. Finally, the mixed effect shows the residual effects of GHG emissions detail of which cannot be captured in this study.

However, to know about the reason behind this positive and negative impact, we have to look at the changes in these factors between the year 1994–1995 and this is given in the Table 5.4.

It is observed from the above table that the GHG emissions intensity for thermal electricity sector is positive. It implies the GHG emissions per unit of output production has increased during this period. This makes the impact of emission intensity positive for thermal electricity sector.

It is observed from Table 5.4 that the use of primary energy per unit of output has increased for thermal electricity, other energy production, and manufacturing sector. Again, these sectors constitute almost 50 % of total GHG emissions in India (INCCA 2010). Therefore, we have observed positive impact of technical coefficient change on GHG emissions.

Table 5.4 shows that the share of agriculture, manufacturing, and primary energy in households consumption basket has declined in the year 2006–2007 as compared to 1994–1995. As the consumption share decreases for these sectors, the gross output production of these sectors will be decreased. As a result, we have observed negative impact on GHG emissions. Due to this decrease in consumption share in these sectors, we have observed negative impact.

Finally, if we take into account only these first four factors as described from Eqs. 5.7a–d, their total impact will not match with the observed total impact between the years because there are some other factors which might have significant impact on GHG emissions. But this needs further study which is outside the preview of this study.

Results obtained in this chapter give valuable insights for the Indian economy in the context of driver analysis in climate change impact study. The Indian economy could improve emission intensity between 1994–1995 to 2006–2007 except for thermal electricity sector. During this same period, we have also seen that the manufacturing sector in India has reduced its electricity requirement per unit of output, and further improvement of this will help to reduce total GHG emissions in India. Indian economy with large population (around 50% without access to electricity (MoEF 2009) is expected to have increase in level of consumption of electricity. So, it is important to understand impact of policy measures towards GHG emissions reduction. This needs further research and we have analyzed selected policies it in the next chapter.

References

Dell M, Benjamin FJ, Benjamin AO (2008) Climate change and economic growth: evidence from the last half century. Working Paper 14132, National Bureau of Economic Research (NBER)

Gupta S, Hall S (1997) Stabilizing energy related CO_2 emissions for India. Energy Econ 19(1):125–150

INCCA (2010) India: greenhouse gas emission 2007, Ministry of Environment and Forests, Government of India

Lin S, Chang TC (1996) Decompositon of CO_2, SO_2 and NOx emissions from energy use of major economic sectors in Taiwan. Energy J 17(1):1–17

MoEF (2009) India's greenhouse gas emission inventory–a report of five modeling studies, Ministry of Environment and Forests, Government of India

Mukhopadhyay K (2001) An empirical analysis of sources of CO2 emission changes in India. Asian J Energy Environ 2(3–4):233–271

Pradhan BK, Sahoo A, Saluja MR (1999) A Social accounting matrix for India 1994–95. Economics and political weekly 34(48)

Roy J (2007) De-linking economic growth from GHG emissions through energy efficiency route—how far are we in India? Bull Energy Effic 7(Annual Issue 2007):52–57

Chapter 6
An Environmental Computable General Equilibrium (CGE) Model for India

The input–output model and social accounting matrix (SAM) models have been widely used in building multisectoral, economy-wide models for development planning and policy analysis (Miller 1985). Models of this type assume an economy that is linear in costs with exogenous demands and fixed prices. These models might be appropriate for short-run policy analysis, but their assumptions do not appear to apply to most real-world economies, which are either pure market economies or market-based economies with an overarching governmental presence. Even in the latter type of mixed economies, for example, India, a great deal of economic activity is not under direct control of the government, but is governed by price signals of the market. In such a decentralized system, myriad intersectoral and intrasectoral substitutions, mostly nonlinear, take place in production, consumption, and distribution in response to price changes. Moreover, there are important feedbacks arising out of interactions among the various commodity and factor markets. A computable general equilibrium (CGE) model is especially designed to capture these essential features of the market. CGE analysis, in comparison to other available techniques, thus captures a wider set of economic impacts from a shock or the implementation of a policy reform, typically, by building and then comparing alternative policy scenarios. Therefore, in this study, we have used a CGE model to evaluate the greenhouse gas (GHG) emissions abatement policies. This model is a single-country model interacting with the rest-of-the-world (ROW). We use this to address the primary concern of our policy makers to formulate national climate change mitigation policies which protect national interest without harming global concerns on climate-related issues.

B. D. Pal et al., *GHG Emissions and Economic Growth*,
India Studies in Business and Economics, DOI 10.1007/978-81-322-1943-9_6,
© Springer India 2015

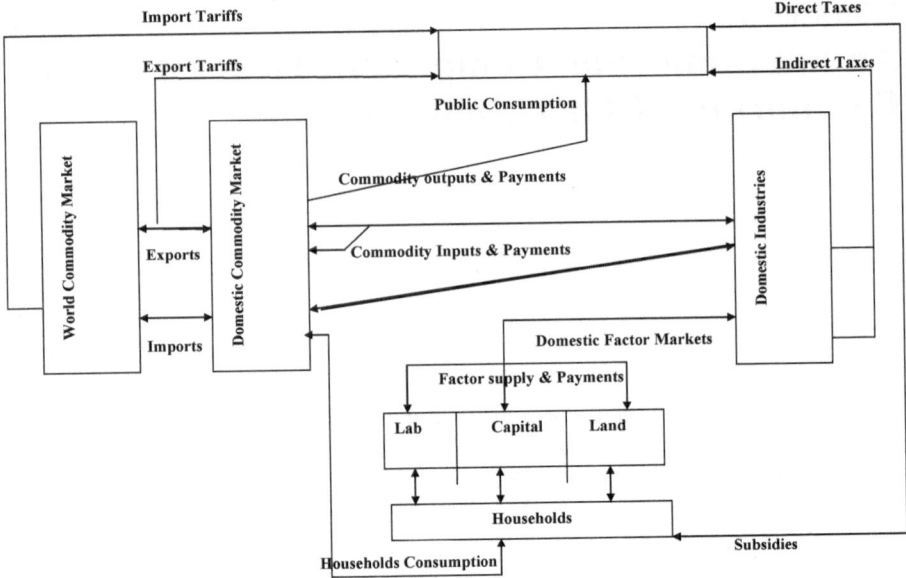

Fig. 6.1 Flow of conventional commodities, factors, payments, and transfer in the economy

6.1   Model Structure

The basic structure of the CGE model adopted here is the one which was prepared jointly by National Council of Applied Economic Research (NCAER) and Jadavpur University (JU) in the year 2009.[1] This CGE model is based on a neoclassical framework that includes institutional features peculiar to the Indian economy. Figure 6.1 depicts the building blocks of CGE model. It is multisectoral and recursively dynamic. The production structure of our model is similar to the one presented by Ghosh (1990) who allows for varying substitution possibilities between different pairs of inputs, which is a crucial feature of the real-world production functions. However, in formulating other elements of the model, such as the linkages between inter-fossil-fuel substitutions, CO_2 emissions, and gross domestic product (GDP), the factor markets and the macroeconomic closure, we follow an eclectic approach taking pieces selectively from Mitra (1994), Fisher-Vanden et al. (1997), and Ojha (2009) and interweave them in an overall consistent framework.

The model includes the interactions of producers, households, the government, and the ROW in response to relative prices given certain initial conditions and exogenously given set of parameters. Producers are guided by profit maximization in perfectly competitive markets, that is, they take factor and output prices (inclusive

---

[1] NCAER (2009) This study was sponsored by Ministry of Environment and Forests, and it was conducted jointly by NCAER and JU. Dr. Sanjib Pohit, Dr. Vijay P Ojha, and Mr. Barun Deb Pal of NCAER and Prof. Joyashree Roy of JU were the core research team members. The report of this project is available at NCAER library on request.

of any taxes) as given and generate demands for factors so as to minimize per unit costs of output. The factors of production include intermediates, energy inputs, and the primary inputs—labor, capital, and land. Production is organized through a multilevel nested production function which has Cobb–Douglas, constant elasticity of substitution (CES), and translog functions at different levels in the production nest. For households, the initial factor endowments are fixed. They, therefore, supply factors inelastically. Households are classified into five rural and four urban socioeconomic groups. Their disposable incomes are defined as their earned incomes net of their savings and the direct taxes paid by them to the government. Their commodity-wise demands are expressed, for given disposable incomes and market prices, through the Stone–Geary linear expenditure system (LES). Government consumption transfers and direct and indirect tax rates are exogenous policy instruments. The total GHG (carbon dioxide equivalent, CO_2EQ emission of CO_2, N_2O, and CH_4) emissions in the economy are determined on the basis of the inputs of fossil fuels in the production process, the gross outputs produced, and the consumption demands of the households and the government, using fixed emission coefficients. The ROW supplies goods to the economy which are imperfect substitutes for domestic output, makes transfer payments, and demands exports. The standard small-country assumption is made implying that India is a price-taker in import markets and can import as much as it wants without affecting the world price of imports. However, because the imported goods are differentiated from the domestically produced goods, the two varieties are aggregated using a CES function, based on the Armington assumption. For exports, a downward-sloping world demand curve is assumed. On the supply side, a constant elasticity of transformation (CET) function is used to define the output of a given sector as a revenue-maximizing aggregate of goods for the domestic market and goods for the foreign markets. Capital stocks are fixed and intersectorally immobile. Rental price of capital is therefore sector specific, depending upon a sector's demand for capital. Supply–demand balances for all commodities and nonfixed factors clear through adjustment in prices in frictionless markets. The model is Walrasian in character, and hence, it determines only relative prices. The overall price index is chosen to be the numéraire and is, therefore, normalized to unity. With the (domestic) price level and the foreign savings fixed exogenously, the model determines endogenously the nominal exchange rate in the external closure and the level of investment in the domestic macro closure (Robinson 1999). In other words, because the foreign savings is exogenously fixed, the model follows a saving-driven macro closure in which the investment level adjusts to satisfy the saving-investment balance.

6.1.1 Sectoral Disaggregation and SAM

Our CGE model is based on a 35-sector disaggregation of the Indian economy. The foundation of a CGE model is its underlying SAM. A multisectoral CGE model derives its validation from replication of the base-year SAM values. Hence, the sectoral classification of the SAM ought to reflect the nuances of the issue under focus in the CGE model—which in this study is carbon emissions abatement—without the number of sectors being so large as to make it practically too cumbersome for any meaningful interpretation. In this matter, our 35-sector SAM developed in Chap. 2 classifying sectors broadly on the basis of their energy and emission intensities serves well as the bedrock of our CGE model.

6.1.2   The Production Structure

Each production sector has a nested production function, with the structure of nesting being the same across sectors. Each sector produces its gross output, employing capital, labor, and an aggregation of its own and other sectors' inputs known as intermediate inputs. The intermediate inputs are broadly of two kinds—energy and non-energy. The different types of inputs, however, combine through differently specified production functions at the various levels in the production nest whose diagram is shown below (Fig. 6.2).

Aggregate of energy inputs (AEN) is formed through a translog (TL) function which combines the five sources of energy, namely, electricity (AGELEC), coal, natural gas (gas), refined oil (pet), and biomass where AGELEC itself is a linear aggregation of the three main sources of electricity—thermal, hydropower, and nuclear (not shown in the Fig. 6.2). The TL function is used because it allows variable (Allen–Uzawa) elasticity of substitution between different pairs of the aforesaid five sources of energy (Roy 2000). The remaining non-factor inputs are referred to as the aggregate materials, MAT, which represents a fixed-coefficients bundle of inputs from the non-energy sectors. The AEN and the MAT combine into aggregate non-factor inputs, QF, through a CES function for which only one substitution elasticity is required. Further up, the QF, labor, capital, and land (in the case of agricultural sectors only) are coalesced into the domestic gross output using a TL function having different substitution elasticities for different pairs of inputs. It may be noted that, the TL function reduces to the much simpler Cobb–Douglas form in case of unit substitution elasticities between the various input pairs and this is actually the case for a subset of the 35 sectors.

Gross domestic output, QX, itself is an aggregate of its two constituents—domestic sales (QD) and exports (QE) —obtained through a CET function. Finally, at the top end of the production nest, QD, and final imports (QM) into a sector are aggregated into a composite output (QQ) for that sector by making use of a CES Armington aggregation function.

6.1.3   Greenhouse Gas Emissions

GHGs are emitted from the burning of fossil fuel inputs in the production process or in the final consumption of households. Also the GHGs are emitted due to wastewater and municipal solid waste generation by the urban households (see INCCA 2010). In addition to GHG emitted by fuel combustion, there may be GHG

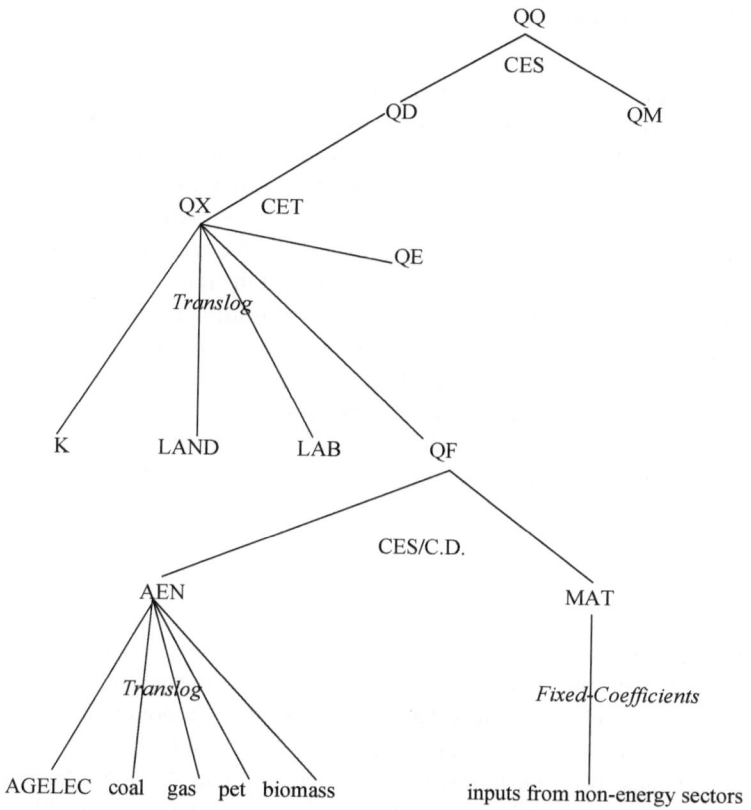

Fig. 6.2 The production nesting. *QQ* composite output, *CES* constant elasticity of substitution, *QD* domestic sales, *QM* final imports, *QX* gross domestic output, *QE* domestic exports, *AEN* aggregate of energy inputs, *MAT* aggregate materials, *CET* constant elasticity of transformation

emanating from the very process of output generation. For example, the cement sector releases CO_2 in the limestone calcination process. Finally, GHG emissions also result from the change in land-use pattern (INCCA 2010). But fossil fuel-based GHG emissions is the highest in India (MoEF 2009). On the other hand, the incorporation of the non-fossil-fuel sources of GHG emissions in our CGE model is a tedious job and also it would make our model more complex. So, to manage within the time-bound program of this study, we have only taken into account fossil fuelled GHG emissions in India. As a result of this, the total GHG emissions obtained for the base year of our CGE model (2006–2007) is somewhat less than that in INCCA report of the year 2010.

We use fixed CO_2, N_2O, and CH_4 emission coefficients to calculate the sector-specific GHG emissions from each of the three sources of carbon emissions. For the total GHG emissions generated in the economy, we first aggregate each type of emissions over the 35 sectors and subsequently convert them into CO_2 equivalent emission (INCCA 2010). Then we aggregate CO_2EQ emission of each type to obtain total GHG emissions of the economy.

6.1.4   Carbon taxes

Our assumption is that carbon taxes are applicable only on the CO_2EQ emitted in the production process (i.e., on the first two sources of carbon emissions), not on the CO_2EQ emitted in the final consumption of households and the government (the third source of carbon emissions). Carbon taxes are based on the proportion of each fuel's carbon content, i.e., rupees per ton of carbon emitted. The carbon tax rate multiplied by a sector's carbon emissions gives the carbon emission tax payments by that sector. Summing across sectors, we get the total carbon tax payments, which is then recycled to the household sector as additional transfer payments by the government. Finally, the producer's cost function is modified to include the carbon taxes so that they induce a substitution in favor of lower carbon-emitting fossil fuels. However, in the baseline (reference) scenario, the carbon tax rate is kept fixed at zero and there are, therefore, neither carbon tax payments nor any incentive for producers to substitute in favor of less carbon-intensive fossil fuels.

6.1.5   Savings

Household savings are determined residually from their respective budget constraints, which state that household income is either allocated to household consumption or to household savings. Government savings is obtained as sum of the tax and tariff revenues, less the value of its consumption and transfers. Government revenue originates from the following five sources: taxes on domestic intermediates, tariffs on imported intermediates, taxes on consumption and investment, taxes on final imports, and income taxes—i.e., taxes on wage, self-employed, and capital (profit) incomes. All taxes (excluding carbon tax) are of the proportional and *ad valorem* type, and all the tax rates are exogenously given. Government expenditure takes place on account of government consumption and transfers to households, both of which are exogenously fixed. The carbon emission tax revenues augment government savings which get channelized into additional investment as the model follows a savings-driven closure rule. Foreign savings in the model is expressed as the excess of payments for intermediate and final imports over the sum of exports earnings, net current transfers, and net factor income from abroad. The latter two are exogenously given values in the model.

6.1.6   Capital Stocks

Sectoral capital stocks are exogenously given at the beginning of a particular period, and are immobile across sectors within that period. Return on capital are thus variable across sectors. However, our model is recursively dynamic, which means that it is run for many periods as a sequence of equilibria. Between two periods there will be additions to capital stocks in each sector because of the investment undertaken in that sector in the previous period. More precisely, sectoral capital stocks for any year t are arrived at by adding the investments by sectors of destination, net of depreciation, in year $t-1$ to the sectoral capital stocks at the beginning of the year $t-1$.

6.1.7 Land and Labor

Total land supply is fixed, but its use can be allocated across the agricultural sectors. Land rental rates are, therefore, uniform across the sectors. Total labor supply is fixed but is sectorally mobile. Market clearing wage rate is determined by equating the sum of sectoral labor demands to the given total labor supply. However, institutional factors cause the sectoral wage rates to deviate from the market clearing wage rate. A sector-specific wage distortion parameter is used to obtain the sectoral wage rates from the economy-wide market-clearing wage rate.

6.1.8 Commodity Market Clearing, Market Closure and Dynamics

Market clearing equilibrium in the commodity markets is ensured by the condition that sectoral supply of composite commodity must equal demand faced by that sector. In the production structure of the model, the domestic gross output of a sector is split into two components—domestic sales and exports—using a CET function. In turn, domestic sales and final imports are amalgamated through an Armington-type CES function to arrive at the sectoral composite commodity supply. On the other hand, the demand for composite commodity consists of intermediate demand, final demand—which in turn is a sum of consumption, investment, and government demands—and change in stocks.

The model is Walrasian in spirit with the sectoral prices being the equilibrating variables for the market-clearing equations. The Walras law holds, and the model is, therefore, homogeneous of degree zero in prices, determining only relative prices. The price index—defined to be a weighted average of sectoral prices—serves, as the numéraire, and is, therefore, fixed at one.

Finally, the model being typically neoclassical in nature follows a savings-driven closure, in which foreign saving is exogenously fixed and investment is endogenous. The model is multiperiod in nature, where the unit of period is 1 year. However, it is not an intertemporal dynamic optimization model; it is only recursively dynamic. That is, it is solved as a sequence of static single-year CGE models where investment in the current year enhances the available capital stock, and depreciation depletes that stock, amounting to net additions (reductions) to sectoral capital stocks between two periods. Hence, prior to solving the CGE model for any given year—other than the base-year—an interim-period sub-model is worked out to update the sectoral capital stocks.

6.1.9 Technological Change

Total factor productivity (TFP) is exogenous in our model. We examined almost all the available empirical evidence, solicited expert opinion, and made reasoned judgments of different baseline GDP scenarios generated for annual TFP growth rates of 2, 3, and 4 % (coupled with different energy efficiency growth rates) before deciding to assume an annual TFP growth of 3 % for our reference scenario.

Energy-saving technological progress is incorporated in our model by making the autonomous energy-efficiency improvement (AEEI) assumption used in other carbon emission abatement models, such as GREEN (Burniaux et al. 1992) and EPPA (Babiker et al. 2001). As in the EPPA and GREEN, we also assume that AEEI occurs in all sectors except the primary energy sectors (coal, crude petroleum, and natural gas) and the refined oil sector. India has embarked on the path towards energy efficiency since 1980, and its record in energy-efficiency improvement in the last two decades is very encouraging. We made reasoned judgments of trial runs of the model for annual AEEI growth rates of 1, 1.2, 1.4, 1.5, and 2.0 (paired with different annual TFP growth rates) found an annual AEEI growth rate of 1.5 % per annum to be most realistic.

Thus, our reference scenario is based on the assumption of 3.0 and 1.5 % annual growth rates of TFP and AEEI, respectively.

6.2 Mathematical Description of the Model

The CGE model is a system of simultaneous, nonlinear equations. The model is square in the sense that the number of equations is equal to the number of variables. In this class of models, this is a necessary (but not sufficient) condition for the existence of a unique solution. In our case also, we have developed a set of equations in such a way that the number of equations is equal to the number of endogenous variables of our model. The set of sectors (industries), parameters, and variables appearing in the equations of the model are described below.

Sets

AS	All sectors (all 35 sectors of 2006–2007 SAM)
DOMS	Sectors with domestic sales (all 35 sectors of 2006–2007 SAM)
FMAT	Factors of production (k-capital, l-labor, la-land, q–aggregated commodity inputs bundle)
GHGS	Greenhouse gases (CO_2-carbon dioxide, N_2O-nitrogen dioxide)
HHSC	Households classes (rh1, rh2, rh3, rh4, rh5, uh1, uh2, uh3, uh4)
PFAC	Primary factors of production (k, l, la)
SIMP	Sectors with imports (pad, wht, cer, cas, anh, frs, fsh, col, oil, gas, fbv, txl, wod, min, pet, chm, pap, fer, irs, alu, omn, mch, rtm, rnm, air, sea, ser)
SNIMP	Sectors without import (cem, hyd, nhy, nuc, bio, gmn, wat, con, rly, hlm)

SEXP	Sectors with exports (pad, wht, cer, cas, anh, frs, fsh, col, gas, fbv, txl, wod, min, pet, chm, pap, fer, irs, alu, cem, omn, mch, rtm, rnm, rly, air, sea, ser)
SNEXP	Sectors without export (oil, hyd, nhy, nuc, bio, gmn, wat, con, hlm)
SMANU	Manufacturing sectors (txl, wod, min, pet, chm, pap, fer, irs, alu, cem, omn, mch, con, gmn, oil)
SNMAN	Non-manufacturing sectors (pad, wht, cer, cas, anh, frs, fsh, fbv, col, gas, bio, hyd, nhy, nuc, wat, rtm, rnm, rly, air, sea, hlm, ser)
SEN	Conventional energy sectors (col, gas, pet, hyd, nhy, nuc, bio)
SMAT	Material input supply sectors (pad, wht, cer, cas, anh, frs, fsh, oil, fbv, txl, wod, min, chm, pap, fer, irs, alu, cem, con, wat, gmn, omn, mch, rly, rtm, rnm, air, sea, hlm, ser)

Endogenous Variables of the Model

AEN_i	Aggregate energy inputs bundle
$AENC_i$	Aggregate energy costs
CD_j	Consumer demand for Armington commodity
$CO2C_i$	Cost for carbon emission
$CO2E_i$	GHG emissions in CO_2 equivalent
$CO2F_i$	Quantity of CO_2 offsets generated in each industry
$CO2N_i$	Net CO_2 emission by each industry
$CO2Q_i$	Quantity of domestic CO_2 quota and offsets purchased by industries
CO2PUB	Quantity of CO_2 emission due to public energy use
CO2PVT	Quantity of CO_2 emission due to private energy use
$CHSTK_j$	Change in stocks
DENei	Quantity of domestic energy used by each industry
$DHIC_h$	Disposable household income
EC_i	CO_2 emission cost
EXR	Exchange rate
$EPN_{e,i}$	Effective price of composite energy inputs
$FINV_i$	Quantity of fixed investment demand for each Armington commodity
GOVI	Government income
GDP	Gross domestic product
HI	Household income
HIC_h	Class-wise households income
$HCD_{h,j}$	Households consumption demand
HCE_h	Hoseholds consumption expenditure
$INT_{j,i}$	Intermediate input demand
INV	Total investment of the economy
$INVD_i$	Investment demand by each industry (sector)
INTei	Intermediate use of Armington energy commodity by industries
MAT_i	Aggregate material inputs bundle
NATCO2	Net national CO_2 emission
PWE_j	World price of exports
PM_j	Domestic price of imports

PE_j	Price of export commodities in domestic market
PDD_j	Price of domestically produced j sold domestically
PQ_j	Price of Armington composite of commodity j in domestic market
PX_j	Producer's price of domestically produced commodity j
$PF_{f,i}$	Return from primary factors
PEN_i	Aggregate energy price
PMA_i	Aggregate material price
PWC	World trade price of CO_2
PC	Price of CO_2
$PUBD_j$	Public consumption demand
$PUBE$	Total public consumption expenditure
QM_j	Total quantity of commodity j imported
QD_j	Quantity of commodity j sold domestically
QQ_j	Quantity of Armington composite of domestic and imported commodity j
QDC_j	Demand for composite commodity j
QX_j	Quantity of domestically produced commodity j
QE_j	Quantity of domestic commodity j exported
$QF_{f,i}$	Quantity of factors used in each industry
RMD_j	Ratio of imported variety to domestic variety of commodity j
RQD_j	Ratio of domestic commodity to composite commodity
RUW	Rural wage
REN	Rental rate of land
TC_i	Total cost for production

Parameters of the Model:

α_j^{ex}	Scaling parameter for CET equation
α_j^{m}	Scaling parameter for Armington CES function
α_{0i}	Scaling parameter of translog energy aggregation function
$\alpha_{e,i}$	Translog energy aggregation parameter
$\alpha_{mat,i}$	Fraction of materials input in total material input of each industry
$\alpha_{q,i}$	Scale parameter of input aggregation function
$b^i_{e,ep}$	Translog Armington energy aggregation parameter
$b_{i,f}$ and $\gamma^i_{f,f'}$	Translog cost function parameter and $\sum_f b_{i,f} = 0$ and $\sum_f b^i_{ff'} = \sum_{f'} b^i_{ff'} = 0$
$\beta_{h,j}$	Beta parameter of LES function
capital0	Initial quantity of capital
cfor	Foreign exchange inflow in the CO_2 account
δ_j^m	Share parameter for j in Armington function
δ_j^{ex}	Share parameter of CET composition function
$\delta_{q,i}$	Share parameter of commodity input aggregation function
dt_h	Households direct tax
$end_{h,f}$	Endowment of primary factors by households classes

exp_j	Quantity of exports when supply price equals to world price
$fdem^{pub}_j$	Post-tax structure of public demand
gsav	Government savings
$\gamma_{h,j}$	Gamma parameter of LES function
κ_i	CO_2 emission permit allocated to industry i
$kap_{i,j}$	Capital composition parameter
λ_i	Depreciation rate
labor0	Initial quantity of labor
μ	Average annual inflation rate
natnlco20	National CO_2 emission
pwm_j	World price of imports
pwc	World trade price of CO_2
pop_h	Population of each households class
prcl	Price level
$\varphi_{e,g}$	Coefficient of GHG emissions by each energy types
r	Interest rate
rew_i	Real wage
ρ_j^{ex}	Elasticity of transformation for exports and domestics
$\rho_{q,i}$	Commodity input aggregation function
ρ_j^m	Elasticity of substitution for j in Armington function
str_h	Share of households in total transfer
sub	Subsidy rate
sr_h	Households savings rate
sctk	Share of total change in stocks in total investment
σ_i	Elasticity between material and energy inputs
σ_j^{ex}	Export demand price elasticity
tm_j	Tariff rate
te_j	Export tax rate
ta_i	Taxes on gross output except export and import tax
ta_i	Taxes on gross output except export and import tax
θ_g	CO_2 equivalent of GHG emissions
tc	Carbon tax
τ	Price level
ϑ_i	Investment share by industry of destination
xch_j	Share of sectoral change in stocks in total change in stocks

Equations of the Model:

1. Domestic price of import commodities

$$PM_j = (1+tm_j).pwm_j.EXR \qquad j\varepsilon\ SIMP$$

2. World price of export commodity

$$PWE_j = PE_j(1+te_j) \qquad j\varepsilon\ SEXP$$

3. Ratio of import to domestic demand

$$RMD_j = \left\{ \left(\frac{\partial_j^m}{1-\partial_j^m} \right) \left(\frac{PDD_j}{PM_j} \right) \right\}^{1/_{1+\rho_j^m}} \qquad j\varepsilon \text{ SIMP} \cap \text{ DOMS}$$

4. Demand for imports

$$QM_j = RMD_j.QD_j \qquad j\varepsilon \text{ SIMP} \cap \text{DOMS}$$

5. Armington equation for composite commodity

$$QQ_j = \alpha_j^m \left\{ \partial_j^m QM_j^{-\rho_j^m} + \left(1-\partial_j^m\right).QD_j^{-\rho_j^m} \right\}^{-1/_{\rho_j^m}} \qquad j\varepsilon \text{ SIMP} \cap \text{DOMS}$$

6. Price of composite of imported variety and domestic variety of commodity j

$$PQ_jQQ_j = QM_j PM_j + QD_j PDD_j \qquad j\varepsilon \text{ AS}$$

7. Composite commodity equation for the case of no imports

$$QQ_j = \alpha_j^m QD_j \qquad j\varepsilon \text{ SNIMP}$$

8. CET equation for exports and domestic

$$QX_j = \alpha_j^{ex} \left\{ \partial_j^{ex} QE_j^{\rho_j^{ex}} + \left(1-\partial_j^{ex}\right) QD_j^{\rho_j^{ex}} \right\}^{1/_{\rho_j^{ex}}} \qquad j\varepsilon \text{ SEXP} \cap \text{DOMS}$$

9. Ratio of exports and domestic demands

$$\left(\frac{QE_j}{QD_j} \right) = \left\{ \left(\frac{PE_j}{PDD_j} \right) \left(\frac{1-\partial_j^{ex}}{\partial_j^{ex}} \right) \right\}^{1/_{\rho_j^{ex}-1}} \qquad j\varepsilon \text{ SEXP} \cap \text{DOMS}$$

10. Price of composite of exports and domestic commodity

$$PX_j QX_j = PE_j QE_j + PDD_j QD_j \qquad j\varepsilon \text{ AS}$$

11. CET equation for no exports

$$QX_j = \alpha_j^{ex} QD_j \qquad j\varepsilon \text{ SNEXP}$$

12. Commodity market balance

$$QQ_i = QDC_i \qquad i\varepsilon \text{ AS}$$

13. Average cost pricing rule for industries

$$QX_i \left(\frac{PX_i}{(1+ta_i)} \right) = TC_i \qquad i\varepsilon \text{ AS}$$

14. Production costs for industries

$$TC_i = \sum_f QF_{f,i} PF_{f,i} + \sum_j INT_{j,i}.PQ_j + CO2C_i \qquad i\varepsilon\,AS$$

$$f\varepsilon\,PFAC$$

15. Quantities of factor inputs (translog production function)

$$QF_{f,i}.PF_{f,i} = (TC_i - CO2C_i).b_{f,i} + \sum_f \gamma^i_{f,f'}.\log(PF_{f,i}) \qquad i\varepsilon\,AS$$

$$f\varepsilon\,FMAT$$

16. Commodity inputs aggregation equation for industries (Cobb-Douglas aggregation function)

$$\ln(QF_{q,i}) = \ln(\alpha_{q,i}).(\partial_{q,i}).\ln(MAT_i) + (1-\partial_{q,i}).\ln(AEN_i) \qquad i\varepsilon\,AS$$

17. Ratio of materials and energy inputs in aggregate commodity inputs

$$MAT_i = AEN_i\left(\left(\frac{PEN_i}{PMA_i}\right)\left(\frac{\partial_{q,i}}{1-\partial_{q,i}}\right)\right)^{\sigma_i} \qquad i\varepsilon\,AS$$

18. Price of aggregate commodity inputs bundle purchased by the user industry

$$PF_{q,i}.QF_{q,i} = (PMA_i.MAT_i + PEN_i.AEN_i) \qquad i\varepsilon\,AS$$

19. Effective cost of aggregate energy inputs bundle

$$AENC_i = \sum_e INT_{e,i}.EPN_{e,i} \qquad i\varepsilon\,AS$$

$$e\varepsilon\,SEN$$

20. Quantity of aggregate energy input bundle

$$\ln(AEN_i) = \ln(AENC_i) - \alpha_{0,i} - \sum_e \alpha_{e,i}.\ln(EPN_{e,i})$$

$$-\frac{1}{2}\sum_e\sum_{ep} b^i_{e,ep}.\ln(EPN_{e,i}).\ln(EPN_{ep,i}) \qquad i\varepsilon\,AS$$

$$e\varepsilon\,SEN$$

21. Quantity of each type of energy used in an industry

$$INT_{e,i}.EPN_{e,i} = \left(\alpha_{e,i} + \sum_{ep} b^i_{e,ep}\ln(EPN_{ep,i})\right) * AENC_i \qquad i\varepsilon\,AS$$

$$e\varepsilon\,SEN$$

22. Aggregation of non-energy commodities as industry intermediate inputs

$$INT_{mat,i} = \alpha_{mat,i}.MAT_i$$

$i\varepsilon$ AS

matε SMAT

23. Price of aggregate material input bundle purchased by industry i.

$$PMA_i.MAT_i = \sum_{mat} PQ_{mat}.INT_{mat,i}$$

$i\varepsilon$ AS

matε SMAT

24. Ratio of domestic commodity to composite commodity

$$RQD_j.QQ_j = QD_j$$

$i\varepsilon$ AS

25. Quantities of domestic and imported energy of each type used as input by an industry.

$$DEN_{e,i} = RQD_j.INT_{e,i}$$

$j\varepsilon$ AS

$e\varepsilon$ SEN

26. GHG emissions by each industry in CO2EQ

$$CO2E_i = \sum_e \sum_g \theta_g.\phi_{e,g}.DEN_{e,i}(1+RMD_e)$$

$i\varepsilon$ AS

$e\varepsilon$ SEN

$g\varepsilon$ GHGS

27. Quantity of CO_2 offsets generated in each industry/sector

$$CO2F_i = \eta_i.QF_{la,i}$$

$i\varepsilon$ AS

28. Net taxable or saleable CO_2 emission by each industry/sector

$$CO2N_i = CO2E_i - CO2F_i - CO2Q_i - \kappa_i$$

$i\varepsilon$ AS

29. Penalty due to positive net CO_2 emission by each industry/sector

$$CO2C_i = CO2N_i.tc + CO2Q_i.PC$$

$i\varepsilon$ AS

30. Effective price of energy input for industry i

$$EPN_{e,i} = PQ_e + PC.RQD_e.(1+RMD_e).\sum_g (\theta_g.\phi_{e,g})$$

$i\varepsilon$ AS

$e\varepsilon$ SEN

$g\varepsilon$ GHGS

31. Effective price of capital for industry i in GE model

$$PF_{k,i} = \tau.(r - \mu) + \lambda_i \qquad i\varepsilon\ AS$$

32. Effective price of labor for industry i in GE model

$$PF_{l,i} = RUW.rew_i \qquad i\varepsilon\ AS$$

33. Effective land rental rate in each industry in GE model

$$PF_{l,i} = REN \qquad i\varepsilon\ AS$$

34. Gross domestic product at factor costs

$$GDP = HI + \sum_j \pi_j^m.QM_j + \sum_j \pi_j^{ex}.QE_j + pwc.QCT.EXR \qquad j\varepsilon\ AS$$

35. Households income by households class

$$HIC_h = \sum_f end_{h,f}.\sum_i PF_{f,i}.QF_{f,i} \qquad i\varepsilon\ AS$$
$$h\varepsilon\ HHSC$$
$$f\varepsilon\ PFAC$$

36. Disposable Household income by households class h

$$DHIC_h = HIC_h - dt_h.(HIC_h - end_{h,la}.\sum_i PF_{la,i}.QF_{la,i}) + str_h.(GDP.sub) \qquad i\varepsilon\ AS$$
$$h\varepsilon\ HHSC$$

37. Net household income

$$HI = \sum_h HIC_h \qquad h\varepsilon\ HHSC$$

38. Net households expenditure

$$HCE_h = DHIC_h.(1 - sr_h) \qquad h\varepsilon\ HHSC$$

39. Total government income

$$GOVI = \sum_h dt_h.\left(HIC_h - end_{h,la}.\sum_i PF_{la,i}.QF_{la,i} \right) + \sum_j QM_j.pwm_j.tm_j.EXR$$
$$+ \sum_j QE_j.PE_j.te_j + \sum_j QA_j.\left(\frac{PX_j}{1+ta_i} \right).pwc.QCT.EXR$$
$$+ PC.QCS + \sum_i tc.CO2N_i + cfor.EXR$$

$$i\varepsilon\ AS$$
$$j\varepsilon\ AS$$
$$h\varepsilon\ HHSC$$

40. Net public consumption expenditure

$$PUBE = GOVI.(1 - gsav) - (sub.GDP)$$

41. Consumer LES demand equations by households class

$$HCD_{h,j} = \left(\gamma_{h,j} + \left(\frac{\beta_{h,j}}{PQ_j} \right) \cdot \left(\left(\frac{HCI_h}{pop_h} \right) - \sum_j PQ_j . \gamma_{h,j} \right) \right) . pop_h \qquad j\varepsilon \text{ AS}$$

$$h\varepsilon \text{ HHSC}$$

42. Consumer demand equation

$$CD_j = \sum_h HCD_{h,j} \qquad j\varepsilon \text{ AS}$$

$$h\varepsilon \text{ HHSC}$$

43. Public demand equation

$$PUBD_j = (fdem_j^{pub} . PUBE) / PQ_j \qquad j\varepsilon \text{ AS}$$

44. Value of gross investment in the economy

$$INV = \sum_h DHIC_h . sr_h + GOVI.gsav + fsav.EXR \qquad h\varepsilon \text{ HHSC}$$

45. Investment demand by each industry

$$INVD_i . PQ_i = \vartheta_i . (1 - sctk).INV \qquad i\varepsilon \text{ AS}$$

46. Quantity of fixed investment demand for each Armington commodity in the economy

$$FINV_j = \left(\left(\sum_i kap_{i,j} . INVD_i \right) \Big/ PQ_j \right) \qquad i\varepsilon \text{ AS}$$

$$j\varepsilon \text{ AS}$$

47. Quantity of change in stock demand for each composite commodity

$$CHSTK_j . PQ_j = xch_j . sctk.INV \qquad j\varepsilon \text{ AS}$$

48. Total demand of composite commodity j in the domestic market

$$QDC_j = PUBD_j + CD_j + INVD_j + CHSTK_j + \sum_i INT_{j,i} \qquad j\varepsilon \text{ AS}$$

49. Export demands for domestic commodities

$$QE_j = \exp_j \cdot \left(\frac{pwm_j}{\left(\frac{pwm_j}{EXR} \right)} \right)^{\sigma_j^{ex}} \qquad j\varepsilon\, \text{SEXP}$$

50. Labor market balance

$$\sum_i QF_{l,i} = labor0 \qquad\qquad i\varepsilon\, \text{AS}$$

51. CO2 emission due to final consumer demands for energy commodities

$$CO2PVT = \sum_e (CD_j.RQD_j.(1+RMD_j)).\sum_g (\theta_g.\phi_{e,g}) \qquad j\varepsilon\, \text{AS}$$

$$e\varepsilon\, \text{SEN}$$

$$g\varepsilon\, \text{GHGS}$$

52. CO2 emission due to final public demands for energy

$$CO2PUB = \sum_e (PUBD_j.RQD_j.(1+RMD_j)).\sum_g (\theta_g.\phi_{e,g}) \qquad j\varepsilon\, \text{AS}$$

$$e\varepsilon\, \text{SEN}$$

$$g\varepsilon\, \text{GHGS}$$

53. Net national CO2 emission

$$NATCO2 = CO2PVT + CO2PUB + \sum_i (CO2E_i - CO2F_i) \qquad i\varepsilon\, \text{AS}$$

54. Domestic CO2 balance in the economy

$$QCT - \sum_i CO2Q_i = 0 \qquad\qquad i\varepsilon\, \text{AS}$$

55. External CO2 balance of the national economy

$$NATCO2 + QCT = natnlco2$$

56. Price normalization equation

$$\sum_j wgt_j.PQ_j = prcl \qquad\qquad j\varepsilon\, \text{AS}$$

Updating Equations of the Interim-period Submodel:

$$\overline{K}_{i(t+1)} = \overline{K}_{it} * (1 - dp_i) + INVDT_i$$

$$LS_{(t+1)} = LS_t (1 - dh_1) + n * \Delta P_t$$

6.3   Data Sources of the Model

The principal data source for estimating the parameters of our CGE model is the 35-sector SAM of the Indian economy for the year 2006–2007, which is the base year for our model. This SAM helps us to estimate the entire shift and the share parameters of our CGE model. These estimates are based on the assumption that the base-year SAM represents a set of "equilibrium" values for that particular year. This assumption of benchmark equilibrium reflected in the SAM is very helpful as it provides a great deal of prior information for parameter estimation by calibration. But the SAM does not provide us with the data for estimating all the parameters of our CGE model. To estimate the parameters not obtainable from the SAM, we have used data from National Accounts Statistics (NAS), Annual Survey of Industries (ASI), Public Finance Statistics of India, and various rounds of National Sample Survey Organization (NSSO). As the required time series and/or cross-sectional data for econometrically estimating the full set of parameters for a CGE model rarely exist, for some parameters such as the elasticity of substitutions, some recourse to drawing independent but relevant econometric estimates from other sources is inevitable. We resorted to this practice only when there was no other option.

We present below a summary account of the exogenous variables and the parameters of the model, and how their estimates were obtained.

Behavioral Parameters

1. Savings rates
2. Demand system parameters
3. Share of aggregate investment earmarked for inventory investment
4. Shares for allocation of total inventory investment into sectoral "Change in Stocks"
5. Share of fixed investment by sector of origin

Fiscal Policy Parameters

1. Tax and tariff rates
2. Subsidy rates
3. Share of public consumption demand by sector of origin

Exogenous Factor Endowments

1. Land endowment in the economy
2. Sectoral capital stocks

The behavioral and fiscal parameters obtained from our 2006–2007 SAM. Total land endowment and the sectoral capital stocks are also available from the SAM.

Exogenous Prices

1. World prices of commodities

The world prices are assumed to be at unity as they serve as the anchor price against which all prices are measured.

Technological Parameters

1. Substitution elasticities in the production functions
2. Shift parameters in the production functions
3. Share parameters in the production functions
4. Emission coefficients

The elasticity parameters describe the curvature of various structural functions. The structural functions used in our CGE model are: TL production functions, Cobb-Douglas production functions, LES demand functions for households, CES import demand functions, and CET export supply functions. We do not estimate elasticity parameters here. Rather, they are taken from Alan et al. (2006). Given the base-year "equilibrium" dataset contained in the SAM, elasticity coefficients of the production, and aggregation functions, their shift and share parameters are calibrated in such a manner that the base-year CGE model solution replicates the SAM values. The emission coefficients are estimated as already explained in Chap. 3.

Time Series of Exogenous Variables

In a recursively dynamic model, the model is run for multiple periods as a sequence of static equilibria. This necessitates estimation of those exogenous variables which change over time. To be specific, they are the following:

1. Foreign savings
2. Population
3. Total labor supply in the economy

To obtain the time series data on foreign savings, we have used the projected growth rate from the macro-econometric model for the Indian economy prepared by NCAER (2006). This study reveals that the capital inflow other than foreign direct investment (FDI) and net invisible will grow by 18% per year between 10 years, i.e., 2005–2006 to 2015–2016. But during 2015–2016 to 2025–2026, this growth rate will fall down to 15% only for net invisible. The FDI growth rate will move around 5–15% in different sectors for the time span 2005–2006 to 2015–2016. During 2015–2016 to 2025–2026, it will fall down to 3–5% for different sectors. After getting the series of these variables, we have computed the series of foreign saving with the help of following relations for the years 2006–2007 to 2030–2031.

Foreign savings (i.e., current account balance, C.A.B) = (Trade balance + Net invisibles).

C.A.B = (Capital account + Monetary movement).
Capital account = (FDI + Capital inflow other than FDI).

Table 6.1 Time Series of Exogenous Variables (Sources: NCAER 2006; CENSUS 2006; NSSO 2006; Author's estimate)

Year	Population	Labor supply	Foreign savings
2006–2007	1.1091	0.4739	64.9050
2007–2008	1.1264	0.4834	82.6673
2008–2009	1.1439	0.4932	101.5202
2009–2010	1.1617	0.5030	123.5338
2010–2011	1.1770	0.5115	149.2539
2011–2012	1.1944	0.5214	179.3221
2012–2013	1.2121	0.5315	214.4927
2013–2014	1.2300	0.5419	255.6534
2014–2015	1.2482	0.5523	303.8481
2015–2016	1.2640	0.5616	360.3056
2016–2017	1.2800	0.5713	415.3726
2017–2018	1.2962	0.5812	478.8987
2018–2019	1.3126	0.5913	552.2272
2019–2020	1.3292	0.6014	636.9174
2020–2021	1.3440	0.6107	734.7805
2021–2022	1.3578	0.6198	847.9200
2022–2023	1.3717	0.6291	978.7800
2023–2024	1.3858	0.6385	1,130.1999
2024–2025	1.4001	0.6479	1,305.4802
2025–2026	1.4130	0.6567	1,508.4567
2026–2027	1.4250	0.6654	1,743.5894
2027–2028	1.4371	0.6742	2,016.0641
2028–2029	1.4493	0.6832	2,331.9118
2029–2030	1.4616	0.6921	2,698.1481

The time series data on population for the time period 2006–2007 to 2025–2026 is available from CENSUS (2006) of India. We have made a projection on the population growth rate to obtain the population data for the time period 2006–2007 to 2030–2031. In Table 6.1, we have shown the time series data on population for the time period 2006–2007 to 2030–2031. This data reveals that the population in India is increasing throughout the time period 2006–2007 to 2030–2031.

To estimate the labor supply for India, we have used the data on Labour Force Participation Rate (LFPR) of India. The National sample Survey Organization (NSSO), in its 61st round report, gives the same for last five years, i.e., 2000–2005. This report says that the usual status LFPR increased by nearly 2 percentage points for males and about 3 percentage points for female during this 5-year time span. We have taken this growth rate as constant throughout the time period 2006–2007 to 2029–2030 to estimate the labor supply of India for that time period. The estimated data on labor supply is also given in the Table 6.1.

References

Alan HS, Roy J, Sathaye J (2006) Estimating energy augmenting technological change in developing country industries Energy Econ 28(5):720–729

Babiker MH, Reilly JM, Mayer M, Eckaus RS, Wing IS, Hyman RC (2001) The MIT emissions prediction and policy analysis (EPPA) model: revisions, sensitivity and comparison of results. Report no 71, MIT Joint Program on the Science and Policy of Global Change. MIT, Cambridge

Burniaux JM, Nicoletti G, Martins JO (1992) GREEN: a global model for quantifying the cost of policies to curb CO2 emissions. OECD Economic Studies, no 19. OECD, Paris

CENSUS (2006) Population projections for India and states 2001–2006. Office of the Registrar General and Census Commissioner, New Delhi

Fisher-Vanden KA, Shukla PR, Edmonds JA, Kim SH, Pitcher HM (1997) Carbon taxes and India. Energy Econ 19:289–325

Ghosh P (1990) Simulating greenhouse gases emissions due to energy use by a computable general equilibrium model of national economy. Ph. D. Dissertation Research, Carnegie Mellon University. (UMI, Ann Arbor, order no DA9107559)

INCCA (2010) India: greenhouse gas emission 2007. Ministry of Environment and Forests—Government of India, New Delhi

Miller RE, Blair PD (1985) Input–output analysis: foundations and extensions. Prentice-Hall, Englewood Cliffs

Mitra PK (1994) Adjustment in oil-importing developing countries: a comparative economic analysis. Cambridge University Press, New York

MoEF (2009) "India's greenhouse gas emission inventory"—a report of five modeling studies, Ministry of Environment and Forests, Government of India.

NCAER (2006) Assessing an alternative medium term growth scenario for the indian economy—project report by Shashaka Bhide, D K Pant, and K A Siddiqui, for Confederation of Indian Industry, New Delhi

NCAER (2009) Climate change impact on the Indian economy-A CGE modeling approach—project report by Sanjib Pohit, Vijay P Ojha, and Barun Deb Pal for Ministry of Environment and Forests, Government of India

NSSO (2006) Household consumption expenditure and employment (2004–05), 61st Round, Government of India, Report No 515

Ojha VP (2009) Carbon emission reduction strategies and poverty alleviation in India. Environ Dev Econ 14(03):323–348

Robinson S et al (1999) From stylized to applied models: building multisector CGE models for policy analysis. N Amer J Econ Finance 10(2):5–38

Roy J (2000) Energy indicator analysis for india. Research project sponsored by OECD/IEA, Paris

Chapter 7
Reference and Policy Scenarios of CGE Model

Typically, a computable general equilibrium (CGE) model is employed to develop, first and foremost, what could be called the "no-policy" or "benchmark" scenario, but is conventionally alluded to as the baseline or business-as-usual (BAU) scenario, or simply reference scenario. Subsequently, it is usually run to generate counterfactual policy scenarios, which are then compared with respect to the reference scenario to derive policy lessons. In accordance with this convention in CGE analysis, we have set a twofold objective for this chapter:

1. To build a reference scenario in the absence of a market-based climate change mitigation policy such as carbon tax
2. To develop policy scenarios when carbon tax is instituted to motivate producers to switch to less carbon-intensive fuel inputs

It needs to be stressed that confining the second objective to an analysis of carbon tax only is not because of any limitation of the model. Our CGE model is generic enough to simulate the impact of a variety of market-based climate policy instruments. A fuller utilization of the potential of our model which would include an investigation of many more market-based climate policy instruments could not be achieved in this study, but is part of our larger comprehensive plan already underway. At this point, therefore, it is pertinent to remind the reader that the present study is best perceived as work in progress rather than a conclusive piece of research. Finally, it may be noted that the reasonable assumptions of annual growth rates of 3% and 1.5% for total-factor productivity (TFP) and American Environmental Energy Inc. autonomous energy efficiency improvement (AEEI), respectively mentioned in the previous chapter apply to both the reference and policy scenarios developed in this chapter.

7.1 Reference Scenario

To develop the abovementioned scenarios, our CGE model has been solved using the General Equilibrium Modeling Systems (GAMS) software. Given the benchmark equilibrium dataset of the 2006–2007 social accounting matrix (SAM), the

B. D. Pal et al., *GHG Emissions and Economic Growth,*
India Studies in Business and Economics, DOI 10.1007/978-81-322-1943-9_7,
© Springer India 2015

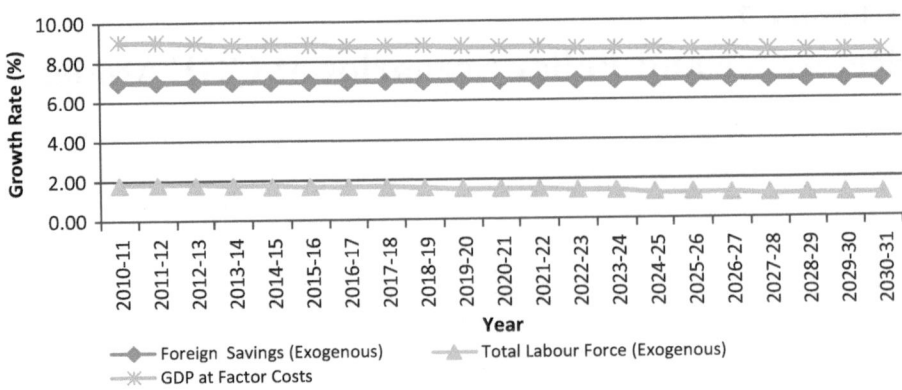

Fig. 7.1 Growth rate of GDP, labor force, and foreign savings. (Source: Author's estimates)

model is calibrated in such a manner that the model-generated solution replicates the base-year data inputs of that year. Thereafter, using the time series of the exogenous variables of the model (given in the previous chapter), we generate a sequence of equilibria for the 24-year period from 2006–2007 to 2030–2031. From the sequence of equilibria, the growth paths of selected (macro) variables of the economy are outlined to describe the reference scenario.

For validation purposes, we have regarded the historical period as the 3-year period from 2006–2007 to 2009–2010 as firm macroeconomic data is available from Central Statistical Organization (CSO) only up to 2009–2010. Since our model runs replicate reasonably well, the actual macroeconomic magnitudes for this period, the model, may be treated as satisfactorily validated.

7.1.1   Gross Domestic Product in the Reference Scenario

In the reference scenario, real gross domestic product (GDP) grows by 9.02% in 2010–2011, after which it grows even faster, i.e., by almost 9% in the following 2 years—2011–2012 and 2012–2013. Thereafter, the real GDP growth rate declines marginally but steadily all through till 2030–2031, reaching therein the level of 8.46% (Fig. 7.1 and Table 7.1). The driving force of GDP growth in our model comes from growth in two main exogenous variables—foreign savings and labor supply—with the former growing faster than the latter. Growth in foreign savings enables an augmentation of investment or capital accumulation. As the capital stock grows faster than labor supply, the relative return on capital declines, inducing a substitution away from labor to capital. This results in an increase in labor productivity (measured as GDP per unit of labor). Hence, growth in labor productivity coupled with the simultaneous growth in labor supply is what provides the main impetus to GDP growth. Furthermore, with there being a constant growth rate in foreign savings, it is evident that the drop in the GDP growth rate follows the decline in the labor force growth rate.

Table 7.1 Macro variables in the reference scenario. (Source: Author's estimate)

Year	Foreign savings (Exogenous) (US$ billion)	Growth rate (Percentage)	Total labor force (Exogenous) Million	Growth rate (Percentage)	GDP at factor costs (Rs. billion)	Growth rate (Percentage)	GDP in PPP (US$ billion)
2010–2011	43.15	7.00	513.04	1.82	46,543.40	9.02	5173.79
2011–2012	46.17	7.00	522.34	1.81	50,720.82	8.98	5638.15
2012–2013	49.40	7.00	531.70	1.79	55,247.20	8.92	6141.31
2013–2014	52.86	7.00	541.07	1.76	60,157.05	8.89	6687.09
2014–2015	56.56	7.00	550.33	1.71	65,482.08	8.85	7279.02
2015–2016	60.52	7.00	559.66	1.70	71,264.63	8.83	7921.81
2016–2017	64.75	7.00	569.04	1.68	77,540.64	8.81	8619.46
2017–2018	69.29	7.00	578.36	1.64	84,350.76	8.78	9376.47
2018–2019	74.14	7.00	587.67	1.61	91,740.14	8.76	10,197.88
2019–2020	79.33	7.00	596.83	1.56	99,745.32	8.73	11,087.74
2020–2021	84.88	7.00	605.99	1.54	108,419.15	8.70	12,051.93
2021–2022	90.82	7.00	615.14	1.51	117,817.23	8.67	13,096.62
2022–2023	97.18	7.00	624.08	1.45	128,002.31	8.64	14,228.80
2023–2024	103.98	7.00	632.94	1.42	139,052.87	8.63	15,457.19
2024–2025	111.26	7.00	641.39	1.34	150,999.07	8.59	16,785.13
2025–2026	119.05	7.00	649.83	1.32	163,908.54	8.55	18,220.16
2026–2027	127.38	7.00	658.07	1.27	177,856.80	8.51	19,770.65
2027–2028	136.30	7.00	666.35	1.26	192,930.32	8.48	21,446.23
2028–2029	145.84	7.00	674.70	1.25	209,275.94	8.47	23,263.22
2029–2030	156.05	7.00	683.05	1.24	226,999.21	8.47	25,233.35
2030–2031	166.97	7.00	691.43	1.23	246,207.32	8.46	27,368.53
Average growth rate		*7.00*		*1.64*		*8.75*	

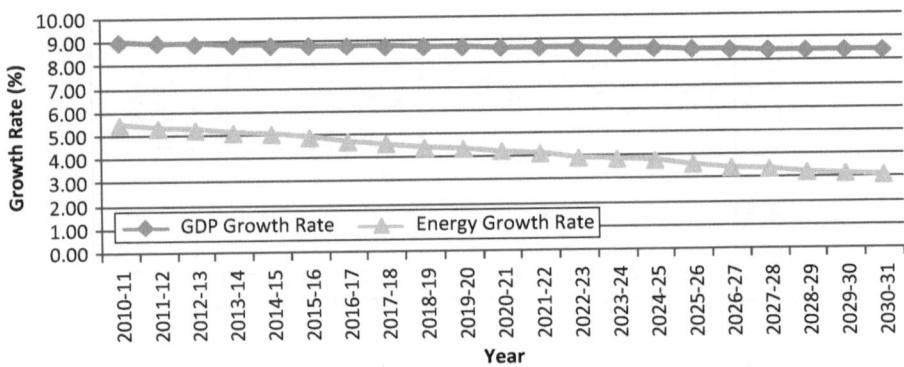

Fig. 7.2 Growth rate of energy use (percentage). (Source: Author's estimate)

Table 7.2 Primary energy use under the assumption of Total Factor Productivity Growth (TFPG)=3 and AEEI=3. (Source: Author's estimate)

Year	GDP in PPP (US$ billion)	GDP growth rate (Percentage)	Primary energy use (Billion tons of oil equivalent)	Energy growth rate (Percentage)	Energy intensity (KgOE/US$ of GDP in PPP)
2010–2011	5173.79	9.02	0.4783	5.50	0.0924
2011–2012	5638.15	8.98	0.5038	5.34	0.0894
2012–2013	6141.31	8.92	0.5303	5.26	0.0863
2013–2014	6687.09	8.89	0.5575	5.12	0.0834
2014–2015	7279.02	8.85	0.5857	5.06	0.0805
2015–2016	7921.81	8.83	0.6144	4.91	0.0776
2016–2017	8619.46	8.81	0.6433	4.71	0.0746
2017–2018	9376.47	8.78	0.6730	4.62	0.0718
2018–2019	10,197.88	8.76	0.7029	4.43	0.0689
2019–2020	11,087.74	8.73	0.7336	4.37	0.0662
2020–2021	12,051.93	8.70	0.7649	4.26	0.0635
2021–2022	13,096.62	8.67	0.7965	4.14	0.0608
2022–2023	14,228.80	8.64	0.8279	3.94	0.0582
2023–2024	15,457.19	8.63	0.8600	3.87	0.0556
2024–2025	16,785.13	8.59	0.8926	3.80	0.0532
2025–2026	18,220.16	8.55	0.9249	3.62	0.0508
2026–2027	19,770.65	8.51	0.9569	3.46	0.0484
2027–2028	21,446.23	8.48	0.9894	3.40	0.0461
2028–2029	23,263.22	8.47	1.0215	3.24	0.0439
2029–2030	25,233.35	8.47	1.0541	3.19	0.0418
2030–2031	27,368.53	8.46	1.0869	3.12	0.0397
Average Growth Rate		*8.75*		*4.66*	

7.1.2   Energy Use in the Reference Scenario

Energy usage grows much less rapidly than GDP in the Indian economy over the 20-year period, 2010–2011 to 2030–2031. The average annual growth rate of primary energy use is 4.66% while that of GDP is 8.75% (Fig. 7.2 and Table 7.2).

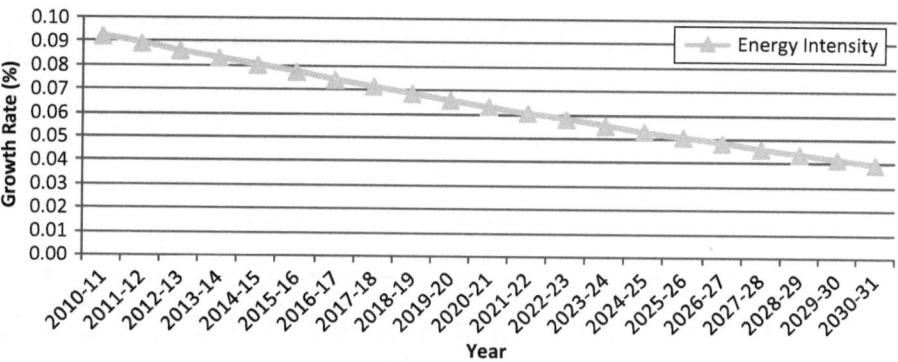

Fig. 7.3 Energy intensity (KgOE/US$ of GDP in PPP). (Source: Author's estimate)

Naturally then, the energy intensity—measured as kg of oil equivalent (KgOE) per US$ of GDP in purchasing power parity (PPP) terms—declines sharply in the 20-year period. From 0.0924 KgOE per US$ of GDP in PPP (KgOE/US$ of GDP in PPP) in 2010–2011, it comes down to 0.0397 (KgOE/US$ of GDP in PPP) in 2030–2031, which is a 65% drop in energy intensity (Fig. 7.3 and Table 7.2). This happens because of the increased substitution of capital in the production process, and the autonomous energy efficiency improvement of 1.5% per annum built into the model.

7.1.3   Carbon Emissions in the Reference Scenario

The fossil-fuel-based carbon dioxide equivalent (CO_2EQ) emissions (i.e., CO_2 plus CO_2EQ of NO_2 and CO_2EQ CH_4) in the period 2011–2030 rise from 1553.46 million tons to 4000.05 million tons at an average rate of 5.25% per year. However, what is significant is that the average annual growth rate of emissions is 3.49% less than the average annual growth rate of GDP. This is explained principally by the sharp decline in the CO_2EQ emissions intensity, which drops from 300.26 grams of CO_2 per US$ of GDP in PPP (grams/US$ of GDP in PPP) in 2010–2011 to 146.16 grams/US$ of GDP in PPP in 2030–2031 (Fig. 7.4 and Table 7.3). The primary cause of the steep descent in carbon emissions intensity is the aforementioned decline in the energy intensity. Fuel switching plays a minimal role.

In assessing India's contribution to global carbon emissions, it is important to look at the per capita carbon dioxide emissions. India's per capita emission (PCE) in 2010–2011 is 1.32 tons. It increases rapidly over the 20-year period and goes up to 2.77 tons by the year 2030 (see Fig. 7.5). However, this level of per capita emissions is considerably less than the global per capita emissions (See MoEF (2009)).

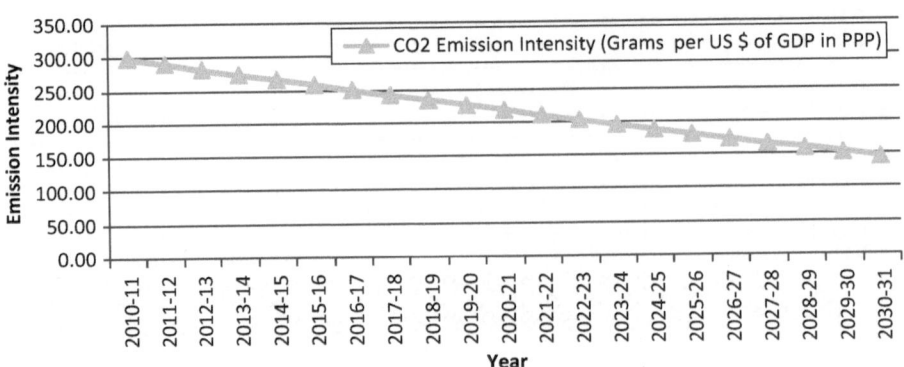

Fig. 7.4 CO$_2$EQ emission intensity (grams/US$ of GDP in PPP). (Source: Author's estimate)

Table 7.3 Total CO$_2$EQ emissions in the reference scenario (Million tons). (Source: Author's estimate)

Year	GDP in PPP (US$ billion)	GDP growth rate (Percentage)	CO$_2$EQ emission (Million tons)	CO$_2$EQ emission growth rate (Percentage)	CO$_2$EQ emission intensity (grams per US$ of GDP in PPP)	Per capita CO$_2$EQ emission (Tons per capita)
2010–2011	5173.79	9.02	1553.46	6.04	300.26	1.32
2011–2012	5638.15	8.98	1645.66	5.94	291.88	1.38
2012–2013	6141.31	8.92	1741.83	5.84	283.62	1.44
2013–2014	6687.09	8.89	1842.67	5.79	275.56	1.51
2014–2015	7279.02	8.85	1947.29	5.68	267.52	1.57
2015–2016	7921.81	8.83	2054.73	5.52	259.38	1.64
2016–2017	8619.46	8.81	2166.64	5.45	251.37	1.71
2017–2018	9376.47	8.78	2282.20	5.33	243.40	1.78
2018–2019	10,197.88	8.76	2399.59	5.14	235.30	1.85
2019–2020	11,087.74	8.73	2521.74	5.09	227.43	1.92
2020–2021	12,051.93	8.70	2648.36	5.02	219.75	2.00
2021–2022	13,096.62	8.67	2775.32	4.79	211.91	2.07
2022–2023	14,228.80	8.64	2905.83	4.70	204.22	2.15
2023–2024	15,457.19	8.63	3037.64	4.54	196.52	2.22
2024–2025	16,785.13	8.59	3171.88	4.42	188.97	2.30
2025–2026	18,220.16	8.55	3309.25	4.33	181.63	2.38
2026–2027	19,770.65	8.51	3446.03	4.13	174.30	2.46
2027–2028	21,446.23	8.48	3582.12	3.95	167.03	2.54
2028–2029	23,263.22	8.47	3720.03	3.85	159.91	2.62
2029–2030	25,233.35	8.47	3858.87	3.73	152.93	2.69
2030–2031	27,368.53	8.46	4000.05	3.66	146.16	2.77
Average growth rate		8.75		5.25		

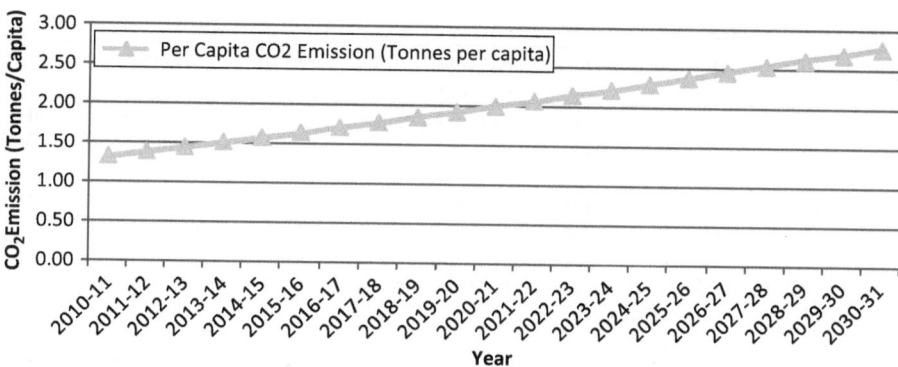

Fig. 7.5 Per capita CO_2EQ emission (tons per capita). (Source: Author's estimate)

7.2 Policy Scenarios

Here we have developed eight alternative policy scenarios for two basic types of domestic carbon tax policy for greenhouse gas (GHG) emissions mitigation: (a) domestic carbon tax enforced with revenue positivity in such a manner that the additional revenue contributes towards an expansion of investment, and (b) domestic carbon tax implemented with revenue neutrality such that there is no additional revenue as the gain in revenue from the carbon taxes is neutralized by a reduction in direct taxes.

For the domestic carbon tax policy with revenue positivity, we have four policy scenarios: scenarios 1a, 1b, 1c, and 1d, dealing with a carbon tax of US\$ 10, US\$ 20, US\$ 40, and US\$ 80 per ton of CO_2EQ emitted, respectively. Likewise, for the domestic carbon tax policy with revenue neutrality, we have four policy scenarios: scenarios 2a, 2b, 2c, and 2d, concerned with US\$ 10, US\$ 20, US\$ 40, and US\$ 80 per ton of CO_2EQ emitted, respectively. So, in effect, there are eight scenarios for the domestic carbon tax policy. The eight policy scenarios are described in Table 7.4.

The adjustment mechanism at work in the policy scenarios needs to be discussed. A carbon tax leads to price increases for each of the fossil fuels—coal, refined oil, and natural gas, but the degree of price increases across these fuels varies because of their differing carbon contents. The price increase is utmost for coal, because coal has the highest carbon content, and least for natural gas which has the lowest carbon content. Producers, motivated by cost minimization, react by substituting refined oil and natural gas for coal as a source of energy. Simultaneously, increased energy prices induce a reduction in overall energy use. Carbon emissions are reduced owing to both fuel switching and overall reduction in fuel input usage. Inter-fossil-fuel substitutions elasticities being low, the fuel reducing impact usually overwhelms the fuel switching impact, causing a deceleration in GDP growth.

Table 7.4 The policy scenarios

	Policy instrument	Revenue neutral/revenue positive	Revenue adjustments
Policy scenario 1a (SIM 1a)	US$ 10 carbon tax imposed per ton of CO_2EQ emitted	Revenue positive	Extra revenue ends up augmenting investment (as it is a saving driven model)
Policy scenario 1b (SIM 1b)	US$ 20 carbon tax imposed per ton of CO_2EQ emitted	Revenue positive	Extra revenue ends up augmenting investment
Policy scenario 1c (SIM 1c)	US$ 40 carbon tax imposed per ton of CO_2EQ emitted	Revenue positive	Extra revenue ends up augmenting investment
Policy scenario 1d (SIM 1d)	US$ 80 carbon tax imposed per ton of CO_2EQ emitted	Revenue positive	Extra revenue ends up augmenting investment
Policy scenario 2a (SIM 2a)	US$ 10 carbon tax imposed per ton of CO_2EQ emitted	Revenue neutral	The direct taxes are reduced to neutralize the additional revenue garnered through carbon taxes
Policy scenario 2b (SIM 2b)	US$ 20 carbon tax imposed per ton of CO_2EQ emitted	Revenue neutral	The direct taxes are reduced to neutralize the additional revenue garnered through carbon taxes
Policy scenario 2c (SIM 2c)	US$ 40 carbon tax imposed per ton of CO_2EQ emitted	Revenue neutral	The direct taxes are reduced to neutralize the additional revenue garnered through carbon taxes
Policy scenario 2d (SIM 2d)	US$ 80 carbon tax imposed per ton of CO_2EQ emitted	Revenue neutral	The direct taxes are reduced to neutralize the additional revenue garnered through carbon taxes

Typically, the adverse impact of reduced energy use on GDP dwindles over time as energy efficiency improvement together with higher capital intensity in the production process results in a falling energy use per unit of GDP.

For carbon tax with revenue positivity scenario, there is additionally an important impact on domestic investment. The net revenue gains by the government lead to extra government savings in our savings-driven model which, result in the expansion of domestic investment. The latter expectedly has an augmenting impact on the GDP, which, however, only mitigates but does not reverse the depressing effect of the carbon taxes on GDP. Thus, the diminishing impact of carbon taxes on the GDP remains predominant, and the net result is that of a decline in GDP.

In case of carbon tax with revenue neutrality, there are no revenue gains (or losses) for the government, but the reduction in direct taxes brought about results in higher disposable incomes for the households, which (depending upon the marginal propensity to consume) is partly consumed and partly saved. Both of these have a favorable impact on GDP. The additional consumption has a demand generating effect, and the extra savings have a supply augmenting effect. The overwhelming impact, however, still comes from the carbon taxes, and it is detrimental to GDP. The net result is, therefore, again a fall in GDP.

7.2.1   Analysis of Results: Revenue Positive Carbon Tax

Below we summarize the findings of our results of important variables related to climate change policy analysis.

GDP in Policy Scenarios 1a, 1b, 1c, 1d In policy scenario 1a, a revenue-positive carbon tax of US$ 10 per ton of CO_2EQ emitted is imposed on the producers, 2010–2011 onwards. As a consequence, the following Table 7.5 shows that the GDP, as compared to the baseline scenario, falls by 1.41 % in 2010–2011. The magnitude of the decline in GDP, however, diminishes progressively in the later years, and it is only 1.01 % in 2030–2031.

In scenario 1b, a revenue-positive carbon tax of US$ 20 per ton of CO_2EQ emitted is enforced, 2010–2011 onwards. Consequently, GDP falls by 2.55 % in 2010–2011. The extent of the decline in GDP again decreases progressively, and reaches the level of 1.79 % in 2030–2031 (See Table 7.5).

In scenario 1c, revenue-positive carbon tax of US$ 40 per ton of CO_2EQ emitted is enforced, again 2010–2011 onwards the result is that GDP falls by 4.56 % in 2010–2011. However, in the later years, the decline becomes increasingly smaller and ends at only 3.30 % in 2030–2031 (See Table 7.5).

Finally, in scenario 1d, a revenue-positive carbon tax of US$ 80 per ton of CO_2EQ emitted is imposed. The impact is that GDP declines by 8.26 % in 2010–2011. The fall in GDP increasingly reduces till it is only 6.20 % in 2030–2031 (Fig. 7.6 and Table 7.5). It follows that the detrimental impact of a carbon tax on the GDP is significantly adverse an increases, though not commensurately, with the magnitude of the tax. Moreover, the adverse impact of the carbon tax on the GDP alleviates over time as GDP grows.

Energy Use in Policy Scenarios 1a, 1b, 1c, 1d Total primary energy use, relative to the baseline scenario, declines for almost all the years in all the policy scenarios, 1a, 1b, 1c, and 1d. And the order of magnitude of the decreases in energy use is linked to that of the declines in the GDP (Table 7.6). Energy intensity, however, is only marginally impacted. What is interesting is that, for all scenarios in almost all the years, the energy intensity has risen even though by a miniscule amount (Table 7.7).

Carbon Emissions in Policy Scenarios 1a, 1b, 1c, 1d CO_2EQ emissions declined significantly in all the policy scenarios 1a, 1b, 1c, and 1d as compared to the reference scenario (Fig. 7.7 and Table 7.8). But the decline in CO_2EQ emissions in percentage terms in each of the years for all the policy scenarios is smaller compared

Table 7.5 GDP in policy scenarios 1a, 1b, 1c, 1d (US$ billion of GDP in PPP). (Source: Author's estimate)

Year	Reference scenario	SIM 1a	% change	SIM 1b	% change	SIM 1c	% change	SIM 1d	% change
2010–2011	5173.79	5100.67	−1.41	5041.76	−2.55	4937.81	−4.56	4746.42	−8.26
2011–2012	5638.15	5559.58	−1.39	5496.47	−2.51	5384.51	−4.50	5178.20	−8.16
2012–2013	6141.31	6057.01	−1.37	5989.29	−2.48	5868.93	−4.44	5646.69	−8.05
2013–2014	6687.09	6596.69	−1.35	6524.10	−2.44	6394.68	−4.37	6155.46	−7.95
2014–2015	7279.02	7182.05	−1.33	7104.42	−2.40	6965.34	−4.31	6707.78	−7.85
2015–2016	7921.81	7817.87	−1.31	7734.82	−2.36	7585.45	−4.25	7308.36	−7.74
2016–2017	8619.46	8508.06	−1.29	8419.26	−2.32	8258.83	−4.18	7960.94	−7.64
2017–2018	9376.47	9257.15	−1.27	9162.22	−2.29	8989.99	−4.12	8669.85	−7.54
2018–2019	10,197.88	10,070.17	−1.25	9968.75	−2.25	9783.90	−4.06	9439.95	−7.43
2019–2020	11,087.74	10,951.19	−1.23	10,842.79	−2.21	10,644.52	−4.00	10,275.14	−7.33
2020–2021	12,051.93	11,905.95	−1.21	11,790.22	−2.17	11,577.72	−3.93	11,180.97	−7.23
2021–2022	13,096.62	12,940.58	−1.19	12,817.15	−2.13	12,589.57	−3.87	12,163.76	−7.12
2022–2023	14,228.80	14,062.01	−1.17	13,930.52	−2.10	13,686.75	−3.81	13,229.95	−7.02
2023–2024	15,457.19	15,279.11	−1.15	15,138.92	−2.06	14,878.11	−3.75	14,388.06	−6.92
2024–2025	16,785.13	16,595.23	−1.13	16,446.06	−2.02	16,167.05	−3.68	15,641.42	−6.81
2025–2026	18,220.16	18,017.51	−1.11	17,859.04	−1.98	17,560.57	−3.62	16,997.31	−6.71
2026–2027	19,770.65	19,554.55	−1.09	19,386.40	−1.94	19,067.37	−3.56	18,464.19	−6.61
2027–2028	21,446.23	21,216.17	−1.07	21,037.54	−1.91	20,696.80	−3.49	20,051.28	−6.50
2028–2029	23,263.22	23,018.50	−1.05	22,828.70	−1.87	22,465.12	−3.43	21,774.07	−6.40
2029–2030	25,233.35	24,973.16	−1.03	24,771.66	−1.83	24,383.76	−3.37	23,643.84	−6.30
2030–2031	27,368.53	27,091.85	−1.01	26,878.31	−1.79	26,464.05	−3.30	25,672.60	−6.20

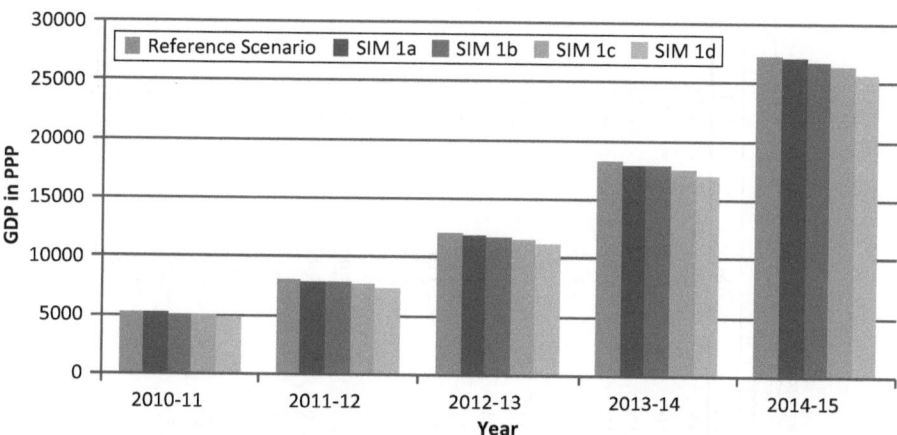

Fig. 7.6 GDP in policy scenarios (US$ billion of GDP in PPP). (Source: Author's estimate)

to the decrease in real GDP (Table 7.9). It follows that the carbon emission intensity falls, even though marginally, for all the years for all the four policy simulations, 1a, 1b, 1c, and 1d (Table 7.10).

As we have seen above in the reference scenario, per capita CO_2EQ emissions are already very low in the Indian economy. The PCE are brought down further by the imposition of a revenue-positive carbon tax, 2010 onwards. In the reference scenario, the PCE in 2020 is 1.32 tons, but it goes up to 2.77 tons in 2030–2031. For a US$ 10 carbon tax (simulation 1a), the PCE is 1.31 tons in 2010, and 2.75 tons in 2030–2031. However, for a US$ 80 carbon tax (simulation 1d), the declines in PCE are of a much higher order of magnitude. For scenario 1d, the PCE is 1.22 in 2010, and 2.60 in 2030–2031 (Fig. 7.8 and Table 7.11).

7.2.2   Analysis of Results: Revenue Neutral Carbon Tax

Below, we illustrate the findings of policy simulations of carbon tax with revenue neutrality. Here, we consider only the major variables related to climate change policy analysis.

GDP in Policy Scenarios 2a, 2b, 2c, 2d In policy scenario 2a, a revenue-neutral carbon tax of US$ 10 per ton of CO_2EQ emitted is imposed on the producers, 2010–2011 onwards. As a consequence, GDP, as compared to the baseline scenario, reduces by 1.36 % in 2010–2011. The extent of the decline in GDP, however, diminishes progressively in the later years, and it is only 0.94 % in 2030–2031.

In scenario 2b, a revenue-neutral carbon tax of US$ 20 per ton of CO_2 emitted is enforced, 2010–2011 onwards. As a result, GDP falls by 2.44 % in 2010–2011. The extent of the decline in GDP again decreases progressively, and reaches the level of 1.62 % in 2030–2031.

Table 7.6 Change in GDP and energy use in policy scenarios 1a, 1b, 1c, 1d (with respect to reference scenario; %). (Source: Authors' estimate)

Year	Percentage change in GDP				Percentage change in energy use			
	SIM1a	SIM1b	SIM1c	SIM1d	SIM1a	SIM1b	SIM1c	SIM1d
2010–2011	−1.41	−2.55	−4.56	−8.26	−0.34	−1.31	−3.33	−6.72
2011–2012	−1.39	−2.51	−4.50	−8.16	−1.13	−1.56	−3.12	−6.63
2012–2013	−1.37	−2.48	−4.44	−8.05	−1.45	−2.16	−4.40	−7.10
2013–2014	−1.35	−2.44	−4.37	−7.95	0.14	−2.52	−3.81	−6.37
2014–2015	−1.33	−2.40	−4.31	−7.85	−0.19	−2.53	−3.90	−6.88
2015–2016	−1.31	−2.36	−4.25	−7.74	−0.31	−2.30	−3.21	−6.93
2016–2017	−1.29	−2.32	−4.18	−7.64	−1.21	−1.83	−2.96	−6.47
2017–2018	−1.27	−2.29	−4.12	−7.54	0.03	−1.24	−3.47	−7.00
2018–2019	−1.25	−2.25	−4.06	−7.43	−0.95	−2.27	−4.00	−6.80
2019–2020	−1.23	−2.21	−4.00	−7.33	−0.84	−1.30	−2.91	−6.60
2020–2021	−1.21	−2.17	−3.93	−7.23	−0.93	−1.85	−2.94	−6.91
2021–2022	−1.19	−2.13	−3.87	−7.12	−0.11	−1.88	−3.25	−6.92
2022–2023	−1.17	−2.10	−3.81	−7.02	−0.67	−1.50	−2.92	−5.94
2023–2024	−1.15	−2.06	−3.75	−6.92	−0.62	−1.76	−3.49	−7.03
2024–2025	−1.13	−2.02	−3.68	−6.81	−0.40	−2.01	−3.17	−6.82
2025–2026	−1.11	−1.98	−3.62	−6.71	−0.83	−1.62	−2.82	−6.31
2026–2027	−1.09	−1.94	−3.56	−6.61	−0.96	−1.23	−3.34	−5.86
2027–2028	−1.07	−1.91	−3.49	−6.50	−1.01	−1.57	−3.47	−6.33
2028–2029	−1.05	−1.87	−3.43	−6.40	−0.86	−1.53	−2.58	−6.47
2029–2030	−1.03	−1.83	−3.37	−6.30	−0.69	−1.62	−3.33	−6.22
2030–2031	−1.01	−1.79	−3.30	−6.20	−0.31	−1.48	−2.59	−5.78

Table 7.7 Energy intensity in policy scenarios 1a, 1b, 1c, 1d (KgOE/US$ of GDP in PPP). (Source: Author's estimate)

Year	Reference scenario	SIM1a	SIM1b	SIM1c	SIM1d
2010–2011	0.0924	0.0934	0.0936	0.0936	0.0940
2011–2012	0.0894	0.0896	0.0902	0.0906	0.0908
2012–2013	0.0863	0.0863	0.0866	0.0864	0.0872
2013–2014	0.0834	0.0846	0.0833	0.0839	0.0848
2014–2015	0.0805	0.0814	0.0804	0.0808	0.0813
2015–2016	0.0776	0.0783	0.0776	0.0784	0.0782
2016–2017	0.0746	0.0747	0.0750	0.0756	0.0756
2017–2018	0.0718	0.0727	0.0725	0.0723	0.0722
2018–2019	0.0689	0.0691	0.0689	0.0690	0.0694
2019–2020	0.0662	0.0664	0.0668	0.0669	0.0667
2020–2021	0.0635	0.0636	0.0637	0.0641	0.0637
2021–2022	0.0608	0.0615	0.0610	0.0612	0.0609
2022–2023	0.0582	0.0585	0.0585	0.0587	0.0589
2023–2024	0.0556	0.0559	0.0558	0.0558	0.0556
2024–2025	0.0532	0.0536	0.0532	0.0535	0.0532
2025–2026	0.0508	0.0509	0.0510	0.0512	0.0510
2026–2027	0.0484	0.0485	0.0488	0.0485	0.0488
2027–2028	0.0461	0.0462	0.0463	0.0461	0.0462
2028–2029	0.0439	0.0440	0.0441	0.0443	0.0439
2029–2030	0.0418	0.0419	0.0419	0.0418	0.0418
2030–2031	0.0397	0.0400	0.0398	0.0400	0.0399

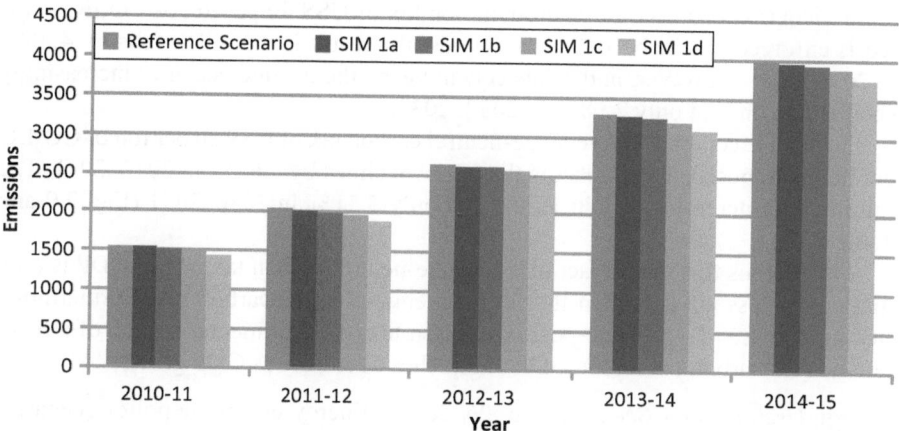

Fig. 7.7 CO_2EQ emission in policy scenarios 1a, 1b, 1c, 1d (million tons). (Source: Author's estimate)

Table 7.8 CO_2EQ emission in policy scenarios 1a, 1b, 1c, 1d. (Million tons) (Source: Author's estimate)

Year	Reference scenario	SIM1a	SIM1b	SIM1c	SIM1d
2010–2011	1553.46	1536.02	1523.99	1484.85	1432.60
2011–2012	1645.66	1627.51	1602.36	1570.36	1512.45
2012–2013	1741.83	1719.33	1706.74	1674.86	1606.53
2013–2014	1842.67	1816.05	1800.82	1768.50	1700.92
2014–2015	1947.29	1930.36	1909.13	1867.25	1793.41
2015–2016	2054.73	2027.35	2002.99	1978.24	1894.57
2016–2017	2166.64	2146.50	2122.46	2083.22	2004.57
2017–2018	2282.20	2251.24	2239.48	2195.20	2110.00
2018–2019	2399.59	2374.72	2342.45	2308.66	2228.81
2019–2020	2521.74	2501.12	2464.36	2430.22	2341.77
2020–2021	2648.36	2617.66	2596.36	2544.70	2463.24
2021–2022	2775.32	2742.08	2717.29	2673.82	2575.81
2022–2023	2905.83	2877.79	2845.03	2804.68	2698.87
2023–2024	3037.64	3010.10	2979.24	2924.10	2837.26
2024–2025	3171.88	3145.67	3107.81	3064.96	2955.50
2025–2026	3309.25	3273.32	3244.29	3192.74	3084.54
2026–2027	3446.03	3410.87	3381.29	3326.46	3217.19
2027–2028	3582.12	3541.94	3514.67	3459.40	3352.24
2028–2029	3720.03	3684.37	3647.94	3587.28	3478.07
2029–2030	3858.87	3826.59	3784.89	3723.94	3616.87
2030–2031	4000.05	3965.49	3931.48	3876.23	3756.20

In scenario 2c, a revenue-neutral carbon tax of US$ 40 per ton of CO_2EQ emitted is enforced; again 2010–2011 onwards, the result is that GDP falls by 4.48% in 2010–2011. However, in the subsequent years, the decline becomes increasingly smaller and ends at only 2.98% in 2030–2031.

Finally, in scenario 2d, a revenue-neutral carbon tax of US$ 80 per ton of CO_2EQ emitted is imposed. The impact is that GDP declines by 7.67% in 2010–2011. The fall in GDP increasingly reduces till it is only 5.31% in 2030–2031 (Fig. 7.9 and Table 7.12).

It is obvious that the impact of a revenue-neutral carbon tax on the GDP is only marginally less adverse than that of a revenue-positive carbon tax. Furthermore, like in the case of a revenue-positive carbon tax, the detrimental impact of a revenue-neutral carbon tax on the GDP diminishes over time as GDP grows.

Energy Use in Policy Scenarios 2a, 2b, 2c, 2d Energy use in the policy scenarios 2a, 2b, 2c, and 2d has changed almost in the same manner as in the policy scenarios 1a, 1b, 1c, and 1d. Total primary energy use, relative to the baseline scenario, declines for almost all the years in the policy scenarios 2a, 2b, 2c, and 2d. And the order of magnitude of the decreases in energy use is linked to that of the declines in the GDP (Table 7.13). For all scenarios in almost all the years, the energy intensity has risen but insignificantly (Table 7.14).

Table 7.9 Change in GDP and CO_2 EQ emission in policy scenarios 1a, 1b, 1c, 1d (with respect to reference scenario). (Source: Author's estimate)

Year	Percentage change in GDP				Percentage change in CO_2 emission			
	SIM1a	SIM1b	SIM1c	SIM1d	SIM1a	SIM1b	SIM1c	SIM1d
2010–2011	−1.41	−2.55	−4.56	−8.26	−1.12	−1.90	−4.42	−7.78
2011–2012	−1.39	−2.51	−4.50	−8.16	−1.10	−2.63	−4.58	−8.09
2012–2013	−1.37	−2.48	−4.44	−8.05	−1.29	−2.01	−3.84	−7.77
2013–2014	−1.35	−2.44	−4.37	−7.95	−1.44	−2.27	−4.03	−7.69
2014–2015	−1.33	−2.40	−4.31	−7.85	−0.87	−1.96	−4.11	−7.90
2015–2016	−1.31	−2.36	−4.25	−7.74	−1.33	−2.52	−3.72	−7.79
2016–2017	−1.29	−2.32	−4.18	−7.64	−0.93	−2.04	−3.85	−7.48
2017–2018	−1.27	−2.29	−4.12	−7.54	−1.36	−1.87	−3.81	−7.54
2018–2019	−1.25	−2.25	−4.06	−7.43	−1.04	−2.38	−3.79	−7.12
2019–2020	−1.23	−2.21	−4.00	−7.33	−0.82	−2.28	−3.63	−7.14
2020–2021	−1.21	−2.17	−3.93	−7.23	−1.16	−1.96	−3.91	−6.99
2021–2022	−1.19	−2.13	−3.87	−7.12	−1.20	−2.09	−3.66	−7.19
2022–2023	−1.17	−2.10	−3.81	−7.02	−0.97	−2.09	−3.48	−7.12
2023–2024	−1.15	−2.06	−3.75	−6.92	−0.91	−1.92	−3.74	−6.60
2024–2025	−1.13	−2.02	−3.68	−6.81	−0.83	−2.02	−3.37	−6.82
2025–2026	−1.11	−1.98	−3.62	−6.71	−1.09	−1.96	−3.52	−6.79
2026–2027	−1.09	−1.94	−3.56	−6.61	−1.02	−1.88	−3.47	−6.64
2027–2028	−1.07	−1.91	−3.49	−6.50	−1.12	−1.88	−3.43	−6.42
2028–2029	−1.05	−1.87	−3.43	−6.40	−0.96	−1.94	−3.57	−6.50
2029–2030	−1.03	−1.83	−3.37	−6.30	−0.84	−1.92	−3.50	−6.27
2030–2031	−1.01	−1.79	−3.30	−6.20	−0.86	−1.71	−3.10	−6.10

Table 7.10 CO_2EQ emission intensity policy scenarios 1a, 1b, 1c, 1d (grams/US$ GDP in PPP). (Source: Author's estimate)

Year	Reference scenario	SIM1a	SIM1b	SIM1c	SIM1d
2010–2011	300.26	301.14	302.27	300.71	301.83
2011–2012	291.88	292.74	291.53	291.64	292.08
2012–2013	283.62	283.86	284.97	285.38	284.51
2013–2014	275.56	275.30	276.03	276.56	276.33
2014–2015	267.52	268.78	268.72	268.08	267.36
2015–2016	259.38	259.32	258.96	260.79	259.23
2016–2017	251.37	252.29	252.10	252.24	251.80
2017–2018	243.40	243.19	244.43	244.18	243.37
2018–2019	235.30	235.82	234.98	235.97	236.10
2019–2020	227.43	228.39	227.28	228.31	227.91
2020–2021	219.75	219.86	220.21	219.79	220.31
2021–2022	211.91	211.90	212.00	212.38	211.76
2022–2023	204.22	204.65	204.23	204.92	204.00
2023–2024	196.52	197.01	196.79	196.54	197.20
2024–2025	188.97	189.55	188.97	189.58	188.95
2025–2026	181.63	181.67	181.66	181.81	181.47
2026–2027	174.30	174.43	174.42	174.46	174.24
2027–2028	167.03	166.95	167.07	167.15	167.18
2028–2029	159.91	160.06	159.80	159.68	159.73
2029–2030	152.93	153.23	152.79	152.72	152.97
2030–2031	146.16	146.37	146.27	146.47	146.31

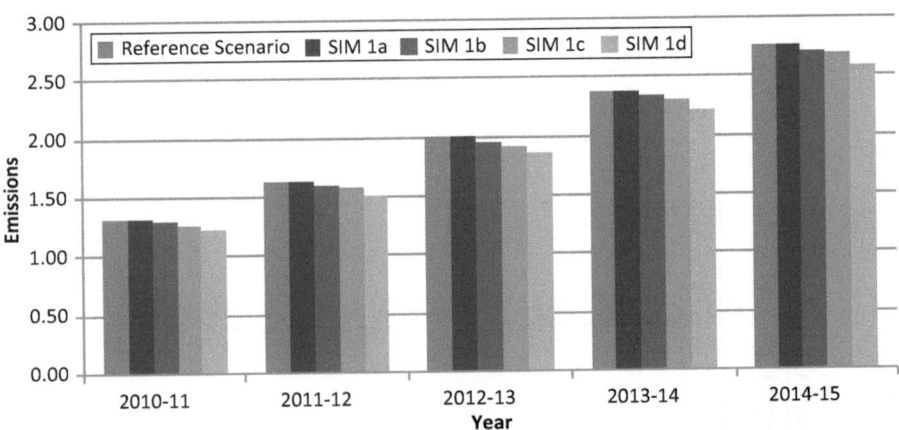

Fig. 7.8 Per capita CO_2EQ emission in policy scenarios 1a, 1b, 1c, 1d (tons/person). (Source: Author's estimate)

Carbon Emissions in Policy Scenarios 2a, 2b, 2c, 2d CO_2EQ emissions decline significantly in all the policy scenarios, 2a, 2b, 2c, and 2d as compared to the reference scenario (Fig. 7.10 and Table 7.15). But the decline in CO_2EQ emissions in percentage terms in each of the years for all the policy scenarios is smaller compared to

Table 7.11 Per capita CO_2EQ emission in policy scenarios 1a, 1b, 1c, 1d. (Million tons per capita) (Source: Author's estimate)

Year	Reference scenario	SIM1a	SIM1b	SIM1c	SIM1d
2010–2011	1.32	1.31	1.30	1.26	1.22
2011–2012	1.38	1.36	1.34	1.32	1.27
2012–2013	1.44	1.42	1.41	1.39	1.33
2013–2014	1.51	1.48	1.47	1.45	1.39
2014–2015	1.57	1.56	1.54	1.51	1.45
2015–2016	1.64	1.62	1.60	1.58	1.51
2016–2017	1.71	1.69	1.67	1.64	1.58
2017–2018	1.78	1.75	1.74	1.71	1.64
2018–2019	1.85	1.83	1.80	1.78	1.72
2019–2020	1.92	1.91	1.88	1.85	1.78
2020–2021	2.00	1.97	1.96	1.92	1.86
2021–2022	2.07	2.05	2.03	2.00	1.92
2022–2023	2.15	2.13	2.10	2.07	2.00
2023–2024	2.22	2.20	2.18	2.14	2.08
2024–2025	2.30	2.28	2.26	2.23	2.15
2025–2026	2.38	2.36	2.34	2.30	2.22
2026–2027	2.46	2.44	2.42	2.38	2.30
2027–2028	2.54	2.51	2.49	2.45	2.38
2028–2029	2.62	2.59	2.57	2.52	2.45
2029–2030	2.69	2.67	2.64	2.60	2.53
2030–2031	2.77	2.75	2.72	2.69	2.60

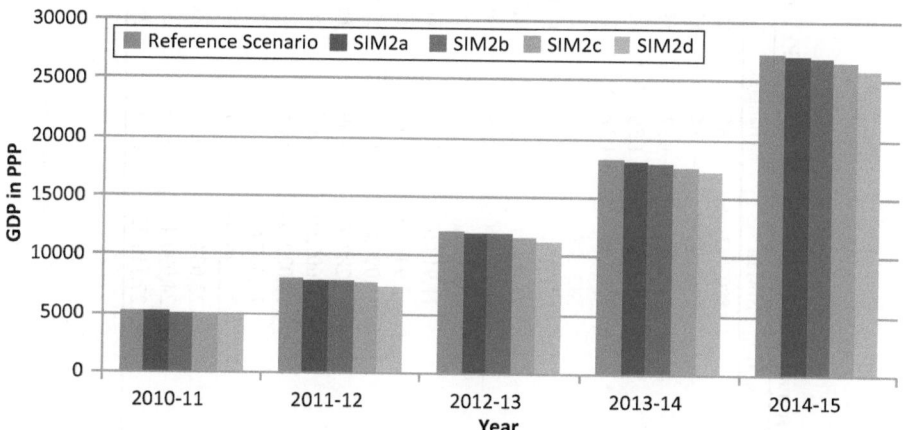

Fig. 7.9 GDP in policy scenarios 2a, 2b, 2c, 2d (US$ billion of GDP in PPP). (Source: Author's estimates)

the decrease in real GDP (Table 7.16). It follows that the carbon emission intensity declines somewhat for all the years for all the four policy simulations, 2a, 2b, 2c, and 2d (Table 7.17).

Table 7.12 GDP in policy scenarios 2a, 2b, 2c, 2d (US$ GDP in PPP). (Source: Author's estimate)

Year	Reference scenario	SIM2a	% change	SIM2b	% change	SIM2c	% change	SIM2d	% change
2010–2011	5173.79	5103.42	−1.36	5047.34	−2.44	4942.01	−4.48	4776.83	−7.67
2011–2012	5638.15	5562.63	−1.34	5502.64	−2.40	5389.79	−4.41	5212.18	−7.56
2012–2013	6141.31	6060.29	−1.32	5996.26	−2.36	5875.42	−4.33	5684.59	−7.44
2013–2014	6687.09	6600.27	−1.30	6531.87	−2.32	6402.61	−4.25	6197.69	−7.32
2014–2015	7279.02	7185.99	−1.28	7113.03	−2.28	6974.80	−4.18	6754.85	−7.20
2015–2016	7921.81	7822.28	−1.26	7744.43	−2.24	7596.73	−4.10	7360.74	−7.08
2016–2017	8619.46	8513.04	−1.23	8429.97	−2.20	8272.17	−4.03	8019.14	−6.96
2017–2018	9376.47	9262.61	−1.21	9174.17	−2.16	9005.63	−3.96	8734.43	−6.85
2018–2019	10,197.88	10,076.10	−1.19	9982.12	−2.12	9802.22	−3.88	9511.54	−6.73
2019–2020	11,087.74	10,957.66	−1.17	10,857.77	−2.07	10,665.85	−3.81	10,354.62	−6.61
2020–2021	12,051.93	11,913.02	−1.15	11,807.02	−2.03	11,602.33	−3.73	11,269.16	−6.49
2021–2022	13,096.62	12,948.53	−1.13	12,835.94	−1.99	12,617.85	−3.66	12,261.51	−6.38
2022–2023	14,228.80	14,070.86	−1.11	13,951.52	−1.95	13,719.44	−3.58	13,338.30	−6.26
2023–2024	15,457.19	15,288.93	−1.09	15,162.15	−1.91	14,915.45	−3.50	14,508.07	−6.14
2024–2025	16,785.13	16,605.89	−1.07	16,471.54	−1.87	16,209.30	−3.43	15,774.28	−6.02
2025–2026	18,220.16	18,029.26	−1.05	17,887.23	−1.83	17,608.71	−3.36	17,144.29	−5.90
2026–2027	19,770.65	19,567.61	−1.03	19,417.33	−1.79	19,121.96	−3.28	18,626.59	−5.79
2027–2028	21,446.23	21,230.60	−1.01	21,071.66	−1.75	20,758.60	−3.21	20,230.62	−5.67
2028–2029	23,263.22	23,034.12	−0.98	22,866.35	−1.71	22,534.55	−3.13	21,972.06	−5.55
2029–2030	25,233.35	24,990.40	−0.96	24,813.43	−1.66	24,461.65	−3.06	23,862.82	−5.43
2030–2031	27,368.53	27,110.72	−0.94	26,924.30	−1.62	26,552.03	−2.98	25,914.19	−5.31

Table 7.13 Change in GDP and energy use in policy scenarios 2a, 2b, 2c, 2d (with respect to reference scenario). (Percentage) (Source: Author's estimates)

Year	Percentage change in GDP				Percentage change in energy use			
	SIM2a	SIM2b	SIM2c	SIM2d	SIM2a	SIM2b	SIM2c	SIM2d
2010–2011	−1.36	−2.44	−4.48	−7.67	−0.03	−1.99	−3.02	−7.39
2011–2012	−1.34	−2.40	−4.41	−7.56	−1.00	−1.87	−3.70	−6.35
2012–2013	−1.32	−2.36	−4.33	−7.44	−0.92	−2.18	−2.91	−6.89
2013–2014	−1.30	−2.32	−4.25	−7.32	−0.57	−2.45	−3.58	−5.82
2014–2015	−1.28	−2.28	−4.18	−7.20	−0.88	−1.82	−2.91	−6.60
2015–2016	−1.26	−2.24	−4.10	−7.08	−0.05	−0.89	−4.02	−5.94
2016–2017	−1.23	−2.20	−4.03	−6.96	−0.90	−1.69	−4.10	−6.51
2017–2018	−1.21	−2.16	−3.96	−6.85	−0.25	−1.53	−3.40	−5.68
2018–2019	−1.19	−2.12	−3.88	−6.73	−1.18	−1.30	−3.30	−6.11
2019–2020	−1.17	−2.07	−3.81	−6.61	−0.87	−1.52	−2.63	−5.75
2020–2021	−1.15	−2.03	−3.73	−6.49	−0.23	−1.82	−3.27	−6.19
2021–2022	−1.13	−1.99	−3.66	−6.38	−1.05	−1.12	−3.23	−5.56
2022–2023	−1.11	−1.95	−3.58	−6.26	−0.08	−1.91	−3.22	−5.46
2023–2024	−1.09	−1.91	−3.50	−6.14	−0.80	−1.72	−2.98	−5.49
2024–2025	−1.07	−1.87	−3.43	−6.02	−0.16	−1.58	−2.95	−5.93
2025–2026	−1.05	−1.83	−3.36	−5.90	−0.86	−1.24	−2.65	−5.22
2026–2027	−1.03	−1.79	−3.28	−5.79	−1.09	−0.94	−3.23	−5.15
2027–2028	−1.01	−1.75	−3.21	−5.67	−0.39	−0.96	−3.14	−5.35
2028–2029	−0.98	−1.71	−3.13	−5.55	−0.71	−1.20	−2.92	−5.23
2029–2030	−0.96	−1.66	−3.06	−5.43	−0.82	−1.08	−2.59	−5.44
2030–2031	−0.94	−1.62	−2.98	−5.31	−0.39	−1.12	−3.01	−5.34

Table 7.14 Energy intensity in policy scenarios 2a, 2b, 2c, 2d. (KgOE/US $ of GDP in PPP) (Source: Author's estimates)

Year	Reference scenario	SIM2a	SIM2b	SIM2c	SIM2d
2010–2011	0.0924	0.0937	0.0929	0.0938	0.0927
2011–2012	0.0894	0.0897	0.0898	0.0900	0.0905
2012–2013	0.0863	0.0867	0.0865	0.0876	0.0869
2013–2014	0.0834	0.0840	0.0833	0.0839	0.0847
2014–2015	0.0805	0.0808	0.0808	0.0815	0.0810
2015–2016	0.0776	0.0785	0.0786	0.0776	0.0785
2016–2017	0.0746	0.0749	0.0750	0.0746	0.0750
2017–2018	0.0718	0.0725	0.0722	0.0722	0.0727
2018–2019	0.0689	0.0689	0.0695	0.0693	0.0694
2019–2020	0.0662	0.0664	0.0665	0.0670	0.0668
2020–2021	0.0635	0.0641	0.0636	0.0638	0.0637
2021–2022	0.0608	0.0609	0.0614	0.0611	0.0613
2022–2023	0.0582	0.0588	0.0582	0.0584	0.0587
2023–2024	0.0556	0.0558	0.0557	0.0559	0.0560
2024–2025	0.0532	0.0537	0.0533	0.0534	0.0532
2025–2026	0.0508	0.0509	0.0511	0.0511	0.0511
2026–2027	0.0484	0.0484	0.0488	0.0484	0.0487
2027–2028	0.0461	0.0464	0.0465	0.0462	0.0463
2028–2029	0.0439	0.0440	0.0441	0.0440	0.0441
2029–2030	0.0418	0.0418	0.0420	0.0420	0.0418
2030–2031	0.0397	0.0399	0.0399	0.0397	0.0397

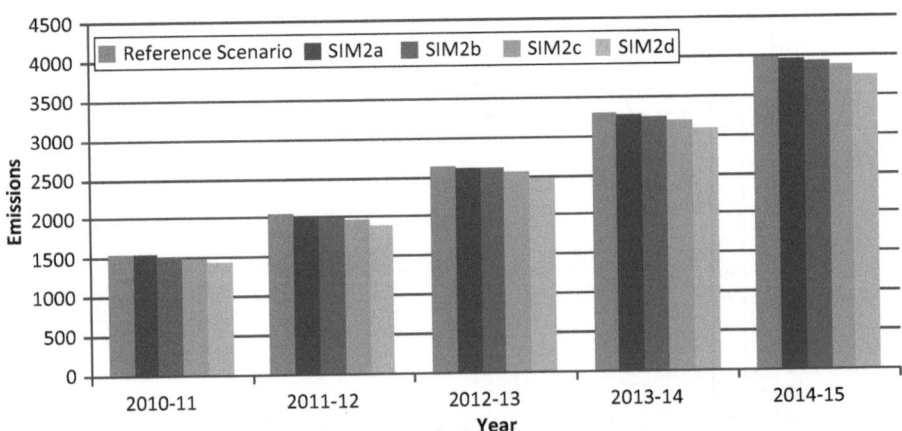

Fig. 7.10 CO_2EQ emission in policy scenarios 2a, 2b, 2c, 2d (million tons). (Source: Author's estimates)

Table 7.15 CO_2EQ emission in policy scenarios 2a, 2b, 2c, 2d (million tons). (Source: Author's estimates)

Year	Reference scenario	SIM2a	SIM2b	SIM2c	SIM2d
2010–2011	1553.46	1536.02	1523.99	1484.85	1432.60
2011–2012	1645.66	1627.51	1602.36	1570.36	1512.45
2012–2013	1741.83	1719.33	1706.74	1674.86	1606.53
2013–2014	1842.67	1816.05	1800.82	1768.50	1700.92
2014–2015	1947.29	1930.36	1909.13	1867.25	1793.41
2015–2016	2054.73	2027.35	2002.99	1978.24	1894.57
2016–2017	2166.64	2146.50	2122.46	2083.22	2004.57
2017–2018	2282.20	2251.24	2239.48	2195.20	2110.00
2018–2019	2399.59	2374.72	2342.45	2308.66	2228.81
2019–2020	2521.74	2501.12	2464.36	2430.22	2341.77
2020–2021	2648.36	2617.66	2596.36	2544.70	2463.24
2021–2022	2775.32	2742.08	2717.29	2673.82	2575.81
2022–2023	2905.83	2877.79	2845.03	2804.68	2698.87
2023–2024	3037.64	3010.10	2979.24	2924.10	2837.26
2024–2025	3171.88	3145.67	3107.81	3064.96	2955.50
2025–2026	3309.25	3273.32	3244.29	3192.74	3084.54
2026–2027	3446.03	3410.87	3381.29	3326.46	3217.19
2027–2028	3582.12	3541.94	3514.67	3459.40	3352.24
2028–2029	3720.03	3684.37	3647.94	3587.28	3478.07
2029–2030	3858.87	3826.59	3784.89	3723.94	3616.87
2030–2031	4000.05	3965.49	3931.48	3876.23	3756.20

In the reference scenario, per capita CO_2 emissions are already very low in the Indian economy. The PCE are brought down further by the imposition of a revenue-neutral carbon tax, 2010 onwards. In the reference scenario, the PCE in 2020 is 1.32 tons, but it goes up to 2.77 tons in 2030–2031. For a US$ 10 carbon tax (simulation 2a), the PCE is 1.31 tons in 2010 and 2.75 tons in 2030–2031. However, for a US$ 80 carbon tax (simulation 1d), the declines in PCE are of a much higher order of magnitude. For scenario 2d, the PCE is 1.22 in 2010, and 2.60 in 2030–2031 (Fig. 7.11 and Table 7.18). In other words, the per capita CO_2 emissions attained under the revenue-neutral carbon tax are nearly the same as those reached in the case of revenue-positive carbon taxes.

In summary, we can say that the India's GHG emissions profile is typically of a low emitter. Even when it traverses on a high GDP growth path of 8%–9% for the next two decades, it ends up with PCE which is substantially lower than the current global average PCE. Also, we can say that a carbon tax—whether it is the revenue-positive type or the revenue-neutral kind—is not an attractive policy option for reducing carbon emissions in India. A carbon tax of whichever kind has a significantly adverse impact on GDP. Moreover, a revenue-neutral carbon tax, in which direct taxes are reduced alongside to maintain the pre-carbon-tax level of revenue, does not appreciably mitigate the GDP loss suffered by the economy in case of the revenue-positive carbon tax.

Table 7.16 Change in GDP and CO_2EQ emission in policy scenarios 2a, 2b, 2c, 2d (with respect to reference scenario). (Source: Author's estimates)

Year	Percentage change in GDP				Percentage change in CO_2EQ emission			
	SIM2a	SIM2b	SIM2c	SIM2d	SIM2a	SIM2b	SIM2c	SIM2d
2010–2011	−1.36	−2.44	−4.48	−7.67	−1.12	−1.90	−4.42	−7.78
2011–2012	−1.34	−2.40	−4.41	−7.56	−1.10	−2.63	−4.58	−8.09
2012–2013	−1.32	−2.36	−4.33	−7.44	−1.29	−2.01	−3.84	−7.77
2013–2014	−1.30	−2.32	−4.25	−7.32	−1.44	−2.27	−4.03	−7.69
2014–2015	−1.28	−2.28	−4.18	−7.20	−0.87	−1.96	−4.11	−7.90
2015–2016	−1.26	−2.24	−4.10	−7.08	−1.33	−2.52	−3.72	−7.79
2016–2017	−1.23	−2.20	−4.03	−6.96	−0.93	−2.04	−3.85	−7.48
2017–2018	−1.21	−2.16	−3.96	−6.85	−1.36	−1.87	−3.81	−7.54
2018–2019	−1.19	−2.12	−3.88	−6.73	−1.04	−2.38	−3.79	−7.12
2019–2020	−1.17	−2.07	−3.81	−6.61	−0.82	−2.28	−3.63	−7.14
2020–2021	−1.15	−2.03	−3.73	−6.49	−1.16	−1.96	−3.91	−6.99
2021–2022	−1.13	−1.99	−3.66	−6.38	−1.20	−2.09	−3.66	−7.19
2022–2023	−1.11	−1.95	−3.58	−6.26	−0.97	−2.09	−3.48	−7.12
2023–2024	−1.09	−1.91	−3.50	−6.14	−0.91	−1.92	−3.74	−6.60
2024–2025	−1.07	−1.87	−3.43	−6.02	−0.83	−2.02	−3.37	−6.82
2025–2026	−1.05	−1.83	−3.36	−5.90	−1.09	−1.96	−3.52	−6.79
2026–2027	−1.03	−1.79	−3.28	−5.79	−1.02	−1.88	−3.47	−6.64
2027–2028	−1.01	−1.75	−3.21	−5.67	−1.12	−1.88	−3.43	−6.42
2028–2029	−0.98	−1.71	−3.13	−5.55	−0.96	−1.94	−3.57	−6.50
2029–2030	−0.96	−1.66	−3.06	−5.43	−0.84	−1.92	−3.50	−6.27
2030–2031	−0.94	−1.62	−2.98	−5.31	−0.86	−1.71	−3.10	−6.10

Table 7.17 CO_2EQ emission intensity in policy scenarios 2a, 2b, 2c, 2d (grams/US$ GDP in PPP). (Source: Author's estimates)

Year	Reference scenario	SIM2a	SIM2b	SIM2c	SIM2d
2010–2011	300.26	301.14	302.27	300.71	301.83
2011–2012	291.88	292.74	291.53	291.64	292.08
2012–2013	283.62	283.86	284.97	285.38	284.51
2013–2014	275.56	275.30	276.03	276.56	276.33
2014–2015	267.52	268.78	268.72	268.08	267.36
2015–2016	259.38	259.32	258.96	260.79	259.23
2016–2017	251.37	252.29	252.10	252.24	251.80
2017–2018	243.40	243.19	244.43	244.18	243.37
2018–2019	235.30	235.82	234.98	235.97	236.10
2019–2020	227.43	228.39	227.28	228.31	227.91
2020–2021	219.75	219.86	220.21	219.79	220.31
2021–2022	211.91	211.90	212.00	212.38	211.76
2022–2023	204.22	204.65	204.23	204.92	204.00
2023–2024	196.52	197.01	196.79	196.54	197.20
2024–2025	188.97	189.55	188.97	189.58	188.95
2025–2026	181.63	181.67	181.66	181.81	181.47
2026–2027	174.30	174.43	174.42	174.46	174.24
2027–2028	167.03	166.95	167.07	167.15	167.18
2028–2029	159.91	160.06	159.80	159.68	159.73
2029–2030	152.93	153.23	152.79	152.72	152.97
2030–2031	146.16	146.37	146.27	146.47	146.31

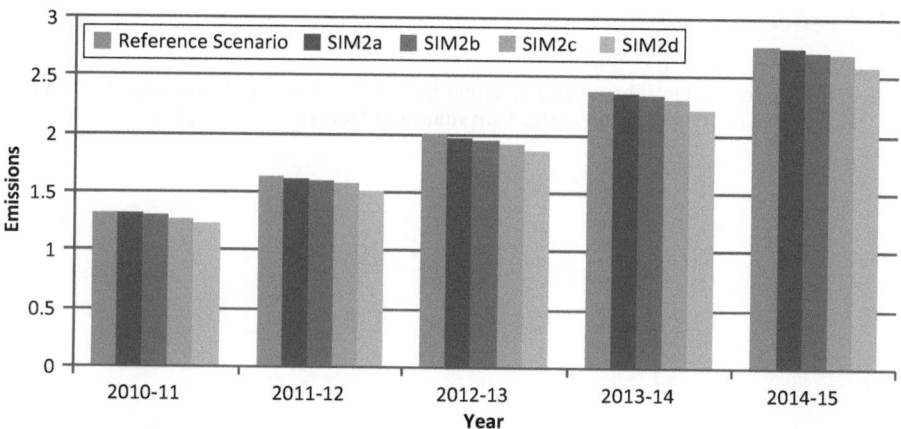

Fig. 7.11 Per capita CO_2EQ emission in policy scenarios 2a, 2b, 2c, 2d (tons/person). (Source: Author's estimates)

Table 7.18 Per capita CO_2EQ emission in policy scenarios 2a, 2b, 2c, 2d (tons per capita). (Source: Author's estimates)

Year	Reference scenario	SIM2a	SIM2b	SIM2c	SIM2d
2010–2011	1.32	1.31	1.30	1.26	1.22
2011–2012	1.38	1.36	1.34	1.32	1.27
2012–2013	1.44	1.42	1.41	1.39	1.33
2013–2014	1.51	1.48	1.47	1.45	1.39
2014–2015	1.57	1.56	1.54	1.51	1.45
2015–2016	1.64	1.62	1.60	1.58	1.51
2016–2017	1.71	1.69	1.67	1.64	1.58
2017–2018	1.78	1.75	1.74	1.71	1.64
2018–2019	1.85	1.83	1.80	1.78	1.72
2019–2020	1.92	1.91	1.88	1.85	1.78
2020–2021	2.00	1.97	1.96	1.92	1.86
2021–2022	2.07	2.05	2.03	2.00	1.92
2022–2023	2.15	2.13	2.10	2.07	2.00
2023–2024	2.22	2.20	2.18	2.14	2.08
2024–2025	2.30	2.28	2.26	2.23	2.15
2025–2026	2.38	2.36	2.34	2.30	2.22
2026–2027	2.46	2.44	2.42	2.38	2.30
2027–2028	2.54	2.51	2.49	2.45	2.38
2028–2029	2.62	2.59	2.57	2.52	2.45
2029–2030	2.69	2.67	2.64	2.60	2.53
2030–2031	2.77	2.75	2.72	2.69	2.60

Reference

MoEF (2009), India's Greenhouse Gas emission Inventory—A report of Five Modeling studies, Ministry of Environment and Forests, Government of India (www.moef.nic.in)

Chapter 8
Policy Message for Mitigating India's Greenhouse Gas (GHG) Emissions

Between the developed and developing countries, it is the latter which are more vulnerable to climate change even though the developed countries have been the primary contributors to greenhouse gas (GHG) emissions till date. Further, it is ironical that the ones who are the most vulnerable to climate change are the least capable to mitigate it, while those who are responsible for causing it in the first place are most unwilling to do their part in resolving it. There is therefore much consensus in the climate change literature on problem statement but none in finding solutions

Among the developing or the late-industrializing economies, Indian economy is a prime example of one which has been an insignificant contributor to the global GHG emissions, but remains highly susceptible to global warming as per Indian Network on Climate Change Assessment (INCCA) report of the year 2010. Notable sectors likely to be affected by climate change are agriculture, forestry, coastal and water resources, and costal land resources due to inundation by a rise in the sea level.

Of late, Indian economy has made significant efforts to reduce its domestic level of GHG emissions. A number of projects have been initiated to reduce GHG emissions, with funding from Global Environment Facility. Most of these projects are on renewable energy sources. In addition to this, policymakers in India are exploring various policy options that would limit carbon emissions. Stronger environmental measures including use of clean fuel and encouraging energy efficiency are a few of them. However, the formulation of appropriate and effective climate change mitigation policies can only follow a rigorous assessment of climate change impact on the economy. Extensive research on climate change impact assessment by different national and international institutes as well as individual environmental economists is ongoing. However, consensus on a comprehensive framework for climate change impact analysis is still emerging.

In the context of climate change impact analysis, most of the studies so far focus on the vulnerable sectors of the economy. However, there are some other sectors which may not be vulnerable sectors but they may be making significant contributions to GHG emissions as well as on economic growth. For example, the manufacturing sector together contributes almost 27 % of total GHG emissions while their share in total gross domestic product (GDP) is almost 18 %. Therefore, growth in

B. D. Pal et al., *GHG Emissions and Economic Growth,*
India Studies in Business and Economics, DOI 10.1007/978-81-322-1943-9_8,
© Springer India 2015

manufacturing sector is likely to have a significantly positive impact on GHG emissions in India.

The reduction in GHG emissions in the Indian economy through decoupling of economic growth and GHG flow is being incorporated in the climate mitigation policy agenda. Capacity to define appropriate policy interventions depends on the economic and technological forces that determine the country's GHG profile. Therefore, we have analyzed in this book the factors that drive India's GHG emissions by decomposing overall emissions into four key determinants namely, intensity effect, intermediate input change effect, final demand mix effect, and final demand level effect.

Indian official statistics, unlike other sectors, do not provide disaggregated time series data on environmental indicators or emissions profile. As a result, Indian researchers are handicapped by data limitation in conducting any quantitative analysis on the linkages between environment and economy. To some extent, this book attempts to reduce this disability by enriching the database by constructing a multisector environmentally extended social accounting matrix (SAM) of India. Unlike earlier SAMs of India, energy sources are disaggregated here. Also, biomass sector, principal source of energy in rural India is segregated in our SAM. Moreover, incorporation of environmental accounts within a SAM is a novel concept in the Indian context although similar accounts for some of the other major countries exist. The methodology for this inclusion is spelt out in detail which would help researchers in this area to construct similar databases for future years.

We have used our environmentally extended SAM to analyze the impact of output growth (and thereby employment growth) on energy use as well as on GHG emissions. It was found that under the fixed technology assumption, the GHG emissions in the economy will increase in the economy if there is expansion in output. Given the technological pattern, the expansion of output requires energy input in the production process, so the demand for energy will increase. Again, due to interdependence between the sectors, there will be direct, indirect, and induced impacts on energy use and GHG emissions, and these are captured with the help of the SAM multipliers.

Expectedly, the aggregate of direct, indirect, and induced impacts of growth in thermal electricity sector on GHG emissions is found to be high. It is also observed that the agricultural sectors and some other sectors having negligible direct impact have significant indirect and induced impact on GHG emissions. This is due to backward and forward linkages between the sectors.

We find that the employment multiplier is high for agriculture and service sector relative to manufacturing sectors. However, the economywide impact on GHG emissions is high for manufacturing sector. On the other hand, the services sector with a significant impact on employment growth does not lead to significant impact on GHG emissions in India.

What are the factors responsible for change in GHG emissions during the period 1994–1995 and 2006–2007? We find from our structural decomposition exercise that changes in emission intensity between the years 1994–1995 and 2006–2007 have negative impacts on GHG emissions. The change in consumption mix during this period also has a negative impact on GHG emissions. But the change in intermediate use per unit of output and the change in level of final demand have a positive impact on GHG emissions.

The use of computable general equilibrium (CGE) model for climate change mitigation policy analysis is widely prevalent in the developed countries. Indian researchers have also employed CGE models for various other policy assessments, but few attempts have been made to use this tool for climate change mitigation analysis. In this book, we have also used a top-down, sequentially dynamic multisectoral Indian environmental CGE model for policy analysis. Our model consists of 35 sectors, with energy-intensive sectors and various types of energy sectors being modeled as separate entities so as to throw light on plausible sector-specific policy interventions. The base year of our model is 2006–2007, same year as the base year of our environmentally extended social accounting matrix (ESAM). In fact, most of the data parameters of the model are drawn from our ESAM. Furthermore, unlike other models which have indiscriminately used parameters drawn from countries other than India, we have used/estimated India-specific parameters for our CGE model.

Our model indicates an average real GDP growth of 8.75 % per annum during the 24-year period, 2006–2007 to 2030–2031 in the business as usual scenario. The model replicates reasonably well the actual annual real GDP growth rates in the so-called historical period 2006–2007 to 2009–2010. Thereafter, the real GDP growth rate declines marginally but steadily all through till 2030–2031 reaching therein the level of 8.46 %.

Energy usage grows much less rapidly than GDP in the Indian economy during the period 2006–2007 to 2030–2031, leading to a sharp decline of 65 % in the energy intensity. The prime driver for this change is the autonomous energy efficiency improvement of 1.5 % per annum built into the model.

Total CO_2 emissions (i.e., CO_2 *plus* CO_2EQ of N_2O) rose from 1004.66 million t in 2006–2007 to 4000.05 million t in 2030–2031 at an average rate of 5.25 % per year. The striking point is the sharp fall in the CO_2 emissions intensity, which drops from 361.54 g of CO_2 per US $ of GDP in PPP (grams/US $ of GDP in purchasing power parity) in 2006–2007 to 146.16 g/US $ of GDP in PPP in 2030–2031. The primary factor for this change is the aforementioned decline in the energy intensity. Fuel switching also plays a significant role.

In assessing India's contribution to global carbon emissions, it is important to look at the per capita CO_2 emissions. India's per capita emission (PCE) in 2006–2007 is 0.95 t. It goes up to 2.77 t by the year 2030. However, this is considerably less than the global PCEs, which is approximately 4.10 t in 2006–2007.

Our CGE model has been used for understanding the impact of economic instruments such as carbon tax on the Indian economy. We have considered in our analysis two types of carbon tax, namely (1) domestic carbon tax enforced with revenue positivity in such a manner that the additional revenue contributes towards an expansion of investment, and (2) domestic carbon tax implemented with revenue neutrality such that there is no additional revenue as the gain in revenue from the carbon taxes is neutralized by a reduction in direct taxes. In each case, the scenarios were done for a range of carbon tax namely, US$ 10, 20, 40, and 80 per tonne of CO_2 emitted.

The principal policy message that emanates from these scenarios is that the detrimental impact of a carbon tax on GDP is significant and increases, though not commensurately, with the magnitude of the tax. Over time, the adverse impact on GDP alleviates. The energy intensity is not significantly affected, nor is the carbon emission intensity much impacted. The decline in carbon emissions, however, is significant but mainly arises due to a marked fall in GDP. This is true, independently of whether one imposes a revenue-positive or a revenue-neutral carbon tax. Thus, the expectations of a double dividend from a revenue-neutral carbon tax are completely belied.

This book presents a modest beginning towards development of detailed Indian database for analysis of environment-related issues. Our CGE analysis is restricted to an investigation of carbon tax only. This is not because of any limitation of the model or the underlying database. Our model is generic enough to simulate the impact a variety of other climate policy instruments. A fuller realization of the potential of the model and the supporting database presented here could not be achieved in one single study. We hope, future research would carry forward our work.

Appendix

Appendix A: Description of 130 Sectors of 2006–2007 IO Tables

1	Paddy	45	Tobacco products	89	Electrical wires and cables
2	Wheat	46	Khadi, cotton textiles(handlooms)	90	Batteries
3	Jowar	47	Cotton textiles	91	Electrical appliances
4	Bajra	48	Woolen textiles	92	Communication equipments
5	Maize	49	Silk textiles	93	Other electrical machinery
6	Gram	50	Art silk, synthetic fibre textiles	94	Electronic equipments(including TV)
7	Pulses	51	Jute, hemp, mesta textiles	95	Ships and boats
8	Sugarcane	52	Carpet weaving	96	Rail equipments
9	Groundnut	53	Readymade garments	97	Motor vehicles
10	Coconut	54	Miscellaneous textile products	98	Motor cycles and scooters
11	Other oilseeds	55	Furniture and fixtures-wooden	99	Bicycles, cycle-rickshaw
12	Jute	56	Wood and wood products	100	Other transport equipments
13	Cotton	57	Paper, paper products and newsprint	101	Watches and clocks
14	Tea	58	Printing and publishing	102	Medical, precision and optical instruments
15	Coffee	59	Leather footwear	103	Jems and jewelry
16	Rubber	60	Leather and leather products	104	Aircraft and spacecraft
17	Tobacco	61	Rubber products	105	Miscellaneous manufacturing
18	Fruits	62	Plastic products	106	Construction
19	Vegetables	63	Petroleum products	107	Electricity

B. D. Pal et al., *GHG Emissions and Economic Growth,*
India Studies in Business and Economics, DOI 10.1007/978-81-322-1943-9,
© Springer India 2015

20	Other crops	64	Coal tar products	108	Water supply
21	Milk and milk products	65	Inorganic heavy chemicals	109	Railway transport services
22	Animal services(agricultural)	66	Organic heavy chemicals	110	Land transport including via pipeline
23	Poultry and eggs	67	Fertilizers	111	Water transport
24	Other livestock products and gobar gas	68	Pesticides	112	Air transport
25	Forestry and logging	69	Paints, varnishes and lacquers	113	Supporting and auxiliary transport activities
26	Fishing	70	Drugs and medicines	114	Storage and warehousing
27	Coal and lignite	71	Soaps, cosmetics and glycerin	115	Communication
28	Natural gas	72	Synthetic fibres, resin	116	Trade
29	Crude petroleum	73	Other chemicals	117	Hotels and restaurants
30	Iron ore	74	Structural clay products	118	Banking
31	Manganese ore	75	Cement	119	Insurance
32	Bauxite	76	Other non-metallic mineral prods.	120	Ownership of dwellings
33	Copper ore	77	Iron, steel and ferro alloys	121	Education and research
34	Other metallic minerals	78	Iron and steel casting and forging	122	Medical and health
35	Lime stone	79	Iron and steel foundries	123	Business services
36	Mica	80	Non-ferrous basic metals	124	Computer and related activities
37	Other non-metallic minerals	81	Hand tools, hardware	125	Legal services
38	Sugar	82	Miscellaneous metal products	126	Real estate activities
39	Khandsari, boora	83	Tractors and agricultural implements	127	Renting of machinery and equipment
40	Hydrogenated oil (vanaspati)	84	Industrial machinery(F & T)	128	O.com, social and personal services
41	Edible oils other than vanaspati	85	Industrial machinery(others)	129	Other services
42	Tea and coffee processing	86	Machine tools	130	Public administration
43	Miscellaneous food products	87	Other non-electrical machinery		
44	Beverages	88	Electrical industrial machinery		

Appendix B: The Social Accounting Matrix of India 2006–2007 (₹ lakh)

	PAD	WHT	CER	CAS	ANH	FRS	FSH	COL	OIL
PAD	3,604,537	56,673	297,765	333	45,486	6	513	0	0
WHT	63,062	2,703,447	377,231	4	13,612	0	6	0	0
CER	53,742	128,243	1,813,875	9	2,555,577	18	7	0	0
CAS	4048	14,766	47,603	798,319	1295	0	0	0	0
ANH	586,772	81,354	954,233	657,037	26,038	1	83	0	1
FRS	132	24	60	0	40	230	0	0	0
FSH	223	571	1182	0	0	0	246,158	0	0
COL	47	35	74	15	0	0	0	11,356	2
OIL	0	0	1	0	202	0	0	0	54,713
GAS	0	0	0	19	2590	0	0	0	0
FBV	45,025	6549	18,793	14	257	0	11,467	0	0
TEX	21,267	21,413	15,603	3338	272,968	393	140,281	88	3
WOD	81	209	433	43	2474	11	6893	29,899	3
MIN	1	3	9	1	236	0	0	0	16
PET	440,637	197,486	428,076	200,894	503	3538	159,517	42,221	96,751
CHM	3040	3257	5886	2217	1292	1043	9975	305,214	68,299
PAP	846	1023	1691	419	19,365	164	0	4436	3
FER	1,551,704	1,351,880	1,533,062	951,908	644	44	219	0	0
CEM	0	0	0	0	194	1	0	0	0
IRS	0	1	7	1	33	1	6856	2	857
ALU	1	3	8	1	1314	19	362	0	1302
OMN	13,173	8489	8675	4004	662	3088	105,725	109,983	157,816
MCH	63,300	74,441	81,915	13,137	2950	633	9	208,316	193,424
NHY	303,163	292,496	154,790	47,095	6681	63	43	105,965	28,621
HYD	56,174	54,198	28,682	8726	919	12	8	19,635	5303
NUC	8532	8232	4356	1325	170	2	1	2982	806
BIO	62,499	9197	108,434	69,230	13,710	776	9	2982	1
WAT	62	53	71	33	16	28	0	1859	0

	PAD	WHT	CER	CAS	ANH	FRS	FSH	COL	OIL
CON	299,358	186,107	241,647	95,450	4927	3305	42	30,402	299,685
LTR	465,228	308,397	416,869	214,952	470,359	10,505	62,115	145,944	54,946
RLY	218,678	54,027	77,352	32,208	28,198	744	2773	8572	3718
AIR	33,234	21,421	15,620	7991	2153	46	2312	697	1147
SEA	1017	1500	3189	362	32,522	25	57	869	543
HLM	0	0	0	0	0	0	0	0	0
SER	848,437	522,990	860,076	350,991	2,020,029	9655	56,472	179,594	192,200
Lab	3,689,555	2,645,224	13,081,859	498,5874	7,939,400	155,015	1,870,083	950,571	880,277
Cap	1,115,629	799,304	3,842,846	140,8910	7,823,192	166,197	1,535,925	2,387,623	2,272,012
Land	2,546,449	1,825,601	9,153,113	350,1862	–	–	–	–	–
RNASE									
RAL									
ROL									
RASE									
ROH									
USE									
USC									
UCL									
UOH									
PVT									
PUB									
GOV	–1,139,709	–1,310,720	–1,108,597	–581,101	31,009	11,379	–202,612	90,587	81514
ITX	53								
CAC									
ROW	31,699	31,699	831,853	283,794	34,943	626,973	15,046	1,059,286	14,830,033
TOT	14,960,001	10,099,591	33,298,342	13,059,415	21,355,984	993,915	4,030,348	5,696,101	19,224,020

	GAS	FBV	TEX	WOD	MIN	PET	CHM	PAP	FER
PAD	0	480,042	41	8	3	0	20,196	1773	340
WHT	0	915,729	74	20	1	0	36,804	137	735
CER	1	3,051,005	7501	503	81	880	308,401	19,754	4427
CAS	2	5,426,983	2,056,049	341	201	1637	728,827	4493	8404
ANH	3	1,601,763	372,582	325	153	308	157,594	882	2331
FRS	4	19,532	290	55,073	37	217	16,076	50,715	69
FSH	0	458,661	40	2	1	0	19,972	172	356
COL	957	29,140	27,400	5525	23,948	252,434	139,313	53,150	32,516
OIL	1376	194	12	300	58	18,673,034	211,974	556	1
GAS	5	3010	36,448	183	1484	886	327,777	2027	381,218
FBV	2	4,300,382	16,609	498	218	1015	479,422	11,745	10,967
TEX	19	106,573	5,501,608	1518	3789	4247	368,217	11,427	9356
WOD	2114	170,905	78,689	14,647	2562	9784	268,310	67,587	24,950
MIN	70	4328	6850	1884	26,313	2631	193,568	4602	162,315
PET	11,260	406,413	407,654	7406	63,514	1,020,947	1,182,041	84,364	791,718
CHM	28,662	1,241,295	2,158,888	38,087	121,069	546,658	15,269,445	421,793	1,275,236
PAP	325	442,279	160,128	19,331	1797	14,496	962,974	617,782	8541
FER	3	61,026	1978	6754	80	3612	225,431	525	577,350
CEM	53	76	370	140	218	3105	7641	70	55
IRS	329	1671	17,261	8151	12,877	4515	215,983	13,309	3104
ALU	116	3941	11,634	3771	49,536	10,981	225,889	7481	5077
OMN	19,255	49,225	196,681	18,950	64,147	20,287	564,638	12,863	9520
MCH	25,725	275,599	558,968	7848	33,712	41,103	484,615	15,348	23,692
NHY	15,172	218,487	616,474	13,510	61,600	294,141	789,216	97,953	79,410
HYD	2811	40,484	114,228	2503	11,414	54,502	146,236	18,150	14,714
NUC	427	6149	17,350	380	1734	8278	22,212	2757	2235
BIO	13	94,233	1534	185,645	130	781	59,890	172,836	313
WAT	133	6695	3641	46	404	209	11,973	115	1729
CON	18,053	333,190	411,837	2372	58,265	192,032	328,759	9861	49,682

	GAS	FBV	TEX	WOD	MIN	PET	CHM	PAP	FER
LTR	17,441	1,665,994	2,120,591	44,524	45,926	120,362	1,709,841	195,932	258,956
RLY	1848	93,918	29,913	4081	14,546	548,377	166,346	24,430	40,166
AIR	153	106,956	7532	3512	1299	15,347	31,377	5648	13,136
SEA	114	19,230	37,301	242	278	3792	52,835	5044	1304
HLM	0	0	0	0	0	0	0	0	0
SER	27,914	5,985,966	4,626,498	126,727	171,480	986,547	4,137,078	325,955	592,151
Lab	422,077	2,468,376	3,448,211	295,948	832,851	306,368	3,575,263	259,182	334,750
Cap	423,125	3,621,031	3,414,704	215,792	2,092,439	4,606,321	7,319,453	423,864	931,988
Land									
RNASE									
RAL									
ROL									
RASE									
ROH									
USE									
USC									
UCL									
UOH									
PVT									
PUB									
GOV									
ITX	13,229	768,166	483,529	20,759	61,959	1,554,145	2,787,084	233,237	301,979
CAC									
ROW	751,027	2,686,079	1,315,523	78,402	8,023,533	3,604,127	8,735,582	820,494	548,325
TOT	1,783,816	37,164,725	28,266,621	1,185,711	11,783,655	32,908,109	52,288,252	3,998,014	6,503,117

	CEM	IRS	ALU	OMN	MCH	NHY	HYD	NUC	BIO
PAD	0	19	1	1578	281	1309	0	0	5115
WHT	0	30	1	3020	546	2266	0	0	3042
CER	44	378	183	16,627	3852	10,779	0	0	213,747
CAS	111	924	729	34,391	9627	16,922	0	0	2881
ANH	59	1459	1161	45,847	12,678	4697	0	0	6689
FRS	19	643	154	35,230	2603	704	0	0	1832
FSH	0	17	1	1639	343	1226	0	0	223
COL	217,084	2,065,513	457,147	574,505	114,056	1,692,816	0	0	164
OIL	2	20,936	1085	91,313	4544	40,917	0	0	209
GAS	35,930	317,837	26,419	70,721	35,564	325,683	0	0	27
FBV	166	2712	2049	22,239	6140	12,255	0	0	24,024
TEX	9504	11,516	4827	180,233	196,041	7383	0	0	3233
WOD	28,946	16,799	5632	171,272	227,330	2250	0	0	350
MIN	368,646	651,084	477,031	1,175,277	213,464	0	0	6146	54
PET	126,628	734,155	145,332	863,213	521,420	1,654,974	0	0	28,271
CHM	137,084	308,761	232,203	2,018,159	2,645,906	117,745	0	1711	10,884
PAP	37,859	20,696	9488	186,071	248,553	27,143	7055	497	3004
FER	17	5835	198	6348	6177	3234	0	0	6528
CEM	1419	4361	1758	72,661	7136	31	24	0	9
IRS	3193	3,560,402	176,752	3,869,196	6,330,835	18,877	2020	0	147
ALU	1372	3,380,076	935,163	2,161,226	4,271,787	18,924	2551	0	212
OMN	171,932	1,036,459	217,809	7,001,479	4,244,945	361,252	12,119	5426	23,057
MCH	7106	342,501	157,092	2,824,181	12,649,358	797,475	40,206	12,174	5690
NHY	248,364	1,113,522	190,742	874,539	603,835	4,222,060	832	61,371	1508
HYD	46,020	206,328	35,343	162,046	111,887	782,319	154	11,372	280
NUC	6990	31,339	5368	24,613	16,994	118,826	23	1727	42
BIO	176	2237	555	119,407	8788	2418	0	0	7717
WAT	23	633	187	20,829	1912	21,709	0	315	211
CON	9669	103,161	68,902	858,778	1,039,096	353,407	21,809	5453	25,876

	CEM	IRS	ALU	OMN	MCH	NHY	HYD	NUC	BIO
LTR	127,948	611,392	161,034	1,316,451	1,293,680	389,306	11,580	5827	118,111
RLY	147,032	955,566	161,873	524,171	211,815	531,289	12,148	7897	8184
AIR	13,577	51,495	11,825	41,876	12,219	53,022	406	776	635
SEA	655	3886	1032	32,578	33,574	4699	853	81	2670
HLM	0	0	0	0	0	0	0	0	0
SER	383,060	3,200,680	651,886	5,679,499	7,705,456	2,107,244	139,633	32,654	240,078
Lab	251,382	2,243,925	786,206	5,060,576	3,681,199	320,002	1,087,917	25,610	1,877,973
Cap	624,803	3,903,004	350,669	7,187,010	4,935,561	2,711,260	1,466,763	24,1362	1,859,180
Land									
RNASE									
RAL									
ROL									
RASE									
ROH									
USE									
USC									
UCL									
UOH									
PVT									
PUB									
GOV									
ITX	111,024	1,029,909	356,922	2,600,023	4,055,593	-1,769,435	-32,813	834	0
CAC						0	0	0	0
ROW	629,557	3,333,828	7,439,439	30,835,793	11,813,863	0	0	0	0
TOT	3,747,403	29,274,019	13,074,201	76,764,616	67,278,659	14,966,987	2,773,280	421,233	4,481,857

	WAT	CON	LTR	RLY	AIR	SEA	HLM	SER	Lab
PAD	49	113	18	0	0	111	3612	718,905	
WHT	93	80	1300	0	0	117	4530	397,700	
CER	388	1,061,416	953,328	0	0	712	13,580	2,341,984	
CAS	611	800	0	0	2	0	0	146,536	
ANH	178	601,652	0	0	0	0	10,338	1,285,048	
FRS	41	83,419	0	0	0	0	0	2992	
FSH	50	43	0	9	0	0	0	17,098	
COL	101	1970	0	4164	0	0	0	20,790	
OIL	619	23	0	0	11	0	0	13,295	
GAS	163	325	0	0	0	0	0	6777	
FBV	543	424	18,874	0	0	837	0	2,467,947	
TEX	112	93,123	79,766	910	50	941	17,421	152,856	
WOD	99	919,489	1068	97	0	0	0	23,867	
MIN	10	3,729,164	0	0	0	0	0	27,148	
PET	2190	3,355,210	12,200,344	268,248	97,767	43,643	70,748	950,541	
CHM	6332	1,141,838	2,324,087	8258	135,793	183,592	2,391,228	1,138,535	
PAP	1524	36,405	153,476	4697	951	460	14,464	287,397	
FER	1241	11,379	230	6	0	0	0	14,203	
CEM	0	3,969,131	0	0	0	0	0	1738	
IRS	2159	10,273,278	891	409	0	0	0	207,244	
ALU	57	3853	326	0	0	0	0	122,373	
OMN	4979	8,386,495	2,113,947	1,250,215	91,698	103,333	133,745	2,538,691	
MCH	4336	2,366,624	933,154	41,763	5363	12,665	104,722	2,176,907	
NHY	22,460	895,618	28,247	695,572	2400	5157	15,150	647,088	
HYD	4162	165,952	5234	128,885	445	956	2807	119,901	
NUC	632	25,206	795	19,576	68	145	426	18,212	
BIO	140	285,604	4041	29	0	3	58	20,010	
WAT	141,549	177,244	15,485	320	985	18,334	900	109,167	
CON	116,775	3,790,532	642,021	866,250	36,259	36,649	222,340	349,1684	

	WAT	CON	LTR	RLY	AIR	SEA	HLM	SER	Lab
LTR	6722	4,110,131	2,382,395	91,440	60,647	77,416	227,590	3,952,304	
RLY	656	981,304	532,260	568,506	1416	1228	1859	114,131	
AIR	47	92,231	142,044	3437	1177	512	1043	57,923	
SEA	183	25,850	41,715	1673	375	201	438,20	103,162	
HLM	0	0	0	100,676	0	0	0	173,246	
SER	89,571	9,607,585	8,210,058	232,405	112,249	176,900	761,775	19,477,882	
Lab	326,430	22,499,481	9,807,816	2,462,740	263,300	57,8908	4,692,207	71,922,015	
Cap	363,570	9,461,837	7,531,709	1,774,196	197,957	415,166	2,693,040	95,799,645	
Land									13,686,272
RNASE									30,655,386
RAL									9,597,032
ROL									23,604,200
RASE									6,014,659
ROH									17,416,862
USE									6,319,0125
USC									9,330,416
UCL									2,282,323
UOH									
PVT									
PUB									
GOV									
ITX	8260	2,468,171	2,627,536	224,044	44,498	48,704	414,210	790,126	
CAC									
ROW	0	0	478,851	0	0	0	0	7,888,498	
TOT	1,107,034	90,622,999	51,231,016	8,748,525	1,053,410	1,706,690	11,841,614	########	175,777,274

	Cap	Land	RNASE	RAL	ROL	RASE	ROH	USE	USC
PAD			921,172	1,929,082	465,192	2,615,038	686,698	882,091	971,107
WHT			527,070	1,103,765	266,169	1,496,249	392,909	504,706	555,640
CER			1,827,586	3,590,841	947,921	4,944,834	1,532,174	2,474,139	2,817,419
CAS			299,626	627,464	151,311	850,582	223,359	286,913	315,868
ANH			1,203,885	1,541,555	566,288	4,038,074	1,205,631	2,093,656	2,433,331
FRS			56,650	106,912	28,156	168,129	46,044	38,522	42,409
FSH			270,564	566,605	136,635	768,082	201,695	259,085	285,231
COL			2587	4806	1392	7050	2297	3982	4539
OIL			0	0	0	0	0	0	0
GAS			8530	15,845	4591	23,244	7574	13,128	14,966
FBV			2,337,903	4,286,378	1,252,981	6,672,925	2,092,675	3,174,404	3,775,280
TEX			1,086,663	1,857,987	535,414	3,279,454	1,057,461	1,576,595	1,917,142
WOD			5587	5428	2970	18,109	5942	11,599	15,773
MIN			0	0	0	0	0	0	0
PET			345,403	571,748	238,310	1,112,680	650,630	723,282	1,928,089
CHM			419,787	578,193	220,158	1,431,805	465,963	850,403	1,564,865
PAP			27,994	36,969	13,736	96,565	26,651	56,321	93,756
FER			0	0	0	0	0	0	0
CEM			0	0	0	0	0	0	0
IRS			0	0	0	0	0	0	0
ALU			0	0	0	0	0	0	0
OMN			330,690	364,709	172,486	1,016,206	356,694	699,474	968,900
MCH			386,716	426,498	201,709	1,188,372	417,126	817,979	1,133,052
NHY			149,390	277,486	80,407	407,068	132,646	229,909	262,088
HYD			27,681	51,416	14,899	75,427	24,578	42,600	48,563
NUC			4204	7810	2263	11,457	3733	6471	7376
BIO			363,625	686,243	180,725	1,079,176	295,545	247,261	272,214
WAT			9931	18,447	5345	27,062	8818	15,284	17,423
CON			170,198	280,178	90,104	537,936	181,655	295,588	465,387

	Cap	Land	RNASE	RAL	ROL	RASE	ROH	USE	USC
LTR			1,616,071	2,134,167	792,934	5,560,509	153,8523	3,251,313	5,412,380
RLY			77,718	102,633	38,133	288,570	73,988	156,357	260,284
AIR			18,739	24,745	9194	−48,923	17,839	37,698	62,755
SEA			58,125	76,759	28,519	307,632	55,336	116,939	194,666
HLM			534,718	1,300,190	354,222	2,104,424	1,185,892	930,716	1,834,783
SER			4,613,267	6,644,682	2,597,189	15,922,245	6,145,036	10,965,155	21,129,082
Lab									
Cap									
Land									
RNASE	11,301,847								
RAL	99,289								
ROL	591,941								
RASE	28,570,049	17,027,026							
ROH	17,120,543								
USE	19,042,479								
USC	4,147,757								
UCL	1,346,862								
UOH	6,309,825								
PVT	32,937,007								
PUB	9,545,700								
GOV	7,439,300		355,411	0	0	4,142,314	1,419,397	0	2,379,634
ITX			802,814	1,317,718	426,876	2,548,941	867,247	1,406,262	2,238,702
CAC	44,737,987		1,055,4871	5,666,113	2,167,887	18,770,185	5,912,228	10,750,880	25,845,192
ROW									
TOT	183,190,587	17,027,026	29,415,175	36,203,372	11,994,117	81,461,420	27,233,983	42,918,713	79,267,896

	UCL	UOH	PVT	PUB	GOV	ITX	CAC	ROW	TOT
PAD	295,510	156,272			102,487		101,482	595,043	14,960,001
WHT	169,082	89,414			53,035		-115,959	533,924	10,099,591
CER	662,037	460,271			185,878		551,811	742,389	33,298,342
CAS	96,119	50,830			0		709,321	141,521	13,059,415
ANH	405,042	396,006			340,150		438,181	282,915	21,355,984
FRS	12,905	6825			61		4938	212,218	993,915
FSH	86,796	45,899			0		4929	656,848	4,030,348
COL	1797	815			6326		-80,482	16,566	5,696,101
OIL	0	0			0		-6482	112,740	19,224,020
GAS	5924	2688			27,381		11,623	75,561	1,783,816
FBV	892,075	647,357			398,033		1,400,206	2,500,625	37,164,725
TEX	342,432	326,420			328,519		594,522	8,394,477	28,266,621
WOD	1513	2298			107		-1,003,963	45,794	1,185,711
MIN	0	0			0		-130,444	4,862,982	117,83655
PET	211,808	209,213			421,402		-2,900,636	2,787,765	32,908,109
CHM	188,901	230,639			908,608		4,992,972	6,114,416	52,288,252
PAP	11,664	15,059			80,533		94,443	167,706	3,998,014
FER	0	0			213		-58,531	240,268	6,503,117
CEM	0	0			0		-330,620	7135	3,747,403
IRS	0	0			0		1,842,627	2,699,301	29,274,019
ALU	0	0			0		543,543	131,3233	13,074,201
OMN	100,834	150,272			539,350		21,869,319	21,128,635	76,764,616
MCH	117,917	175,731			580,572		30,745,595	6,497,608	67,278,659
NHY	103,738	47,068			529,604		0	0	14,966,987
HYD	19,222	8721			98,132		0	0	2,773,280
NUC	2920	1325			14,905		0	0	421,233
BIO	82,835	43,821			0		0	0	4,481,857
WAT	6896	3129			457,828		0	0	1,107,034
CON	69,489	84,079			738,347		73,456,402	0	90,622,999

	UCL	UOH	PVT	PUB	GOV	ITX	CAC	ROW	TOT
LTR	673,354	869,322			844,476		1,440,554	382,4555	51,231,016
RLY	32,382	41,806			151,856		455,432	958,135	8,748,525
AIR	7807	10,080			20,100		33,413	106,138	1,053,410
SEA	24,218	31,267			45,962		197,583	112,451	1,706,690
HLM	322,525	810,765			2,189,457		0	0	11,841,614
SER	2,307,761	3,926,428			3,304,2548		4,512,415	27,110,384	219,745,567
Lab								−251,300	175,777,274
Cap								−2726500	183,190,587
Land									17,027,026
RNASE					3,307,145			1,119,912	29,415,175
RAL					4,070,341			1,378,356	36,203,372
ROL					1,348,497			456,647	11,994,117
RASE					9,158,699			3,101,446	8,146,1420
ROH					3,061,914			1,036,868	2,723,3983
USE					4,825,346			1,634,026	42,918,713
USC					8,912,081			3,017,933	79,267,896
UCL					1,413,124			478,532	12,568,935
UOH					1,137,160			385,081	10,114,388
PVT				9,545,700	2,163,093				35,100,100
PUB									9,545,700
GOV	3,975,924	634,321	14,434,600				8,762,980	4,669,800	75,025,393
ITX	330,440,287	408,564	20,665,500		594,826	35,574,693		795,880	35,574,693
CAC	1,007,065	227,683			−7,072,703			−641,415	148,137,174
ROW									106,696,599
TOT	12,568,935	10,114,388	35,100,100	9,545,700	75,025,393	35,574,693	148,137,174	106,696,599	—

Appendix C: Environmental Social Accounting Matrix of India 2006–2007 (₹ Lakhs for Monetary Transaction and Others in Physical Units)

	PAD	WHT	CER	CAS	ANH	FRS	FSH	COL	OIL
PAD	3,604,537	56,673	297,765	333	45,486	6	513	0	0
WHT	63,062	2,703,447	377,231	4	13,612	0	6	0	0
CER	53,742	128,243	1,813,875	9	2,555,577	18	7	0	0
CAS	4048	14,766	47,603	798,319	1295	0	0	0	0
ANH	586,772	81,354	954,233	657,037	26,038	1	83	0	1
FRS	132	24	60	0	40	230	0	0	0
FSH	223	571	1182	0	0	0	246,158	0	0
COL	47	35	74	15	202	0	0	11,356	2
OIL	0	0	1	0	2590	0	0	0	54,713
GAS	0	0	0	19	257	0	0	0	0
FBV	45,025	6549	18,793	14	272,968	0	11,467	0	0
TEX	21,267	21,413	15,603	3,338	2474	393	14,0281	88	3
WOD	81	209	433	43	236	11	6893	29,899	3
MIN	1	3	9	1	503	0	0	0	16
PET	440,637	197,486	428,076	200,894	1292	3,538	159,517	42,221	96,751
CHM	3040	3257	5886	2217	19,365	1043	9975	305,214	68,299
PAP	846	1023	1691	419	644	164	0	4436	3
FER	1,551,704	1,351,880	1,533,062	951,908	194	44	219	0	0
CEM	0	0	0	0	33	1	0	0	857
IRS	0	1	7	1	1314	1	6856	2	1302
ALU	1	3	8	1	662	19	362	0	23
OMN	13,173	8489	8675	4004	2950	3088	105,725	109,983	157,816
MCH	63,300	74,441	81,915	13,137	6681	633	9	208,316	193,424
NHY	303,163	292,496	154,790	47,095	919	63	43	105,965	28,621
HYD	56,174	54,198	28,682	8726	170	12	8	19,635	5303

	PAD	WHT	CER	CAS	ANH	FRS	FSH	COL	OIL
NUC	8532	8232	4356	1325	26	2	1	2982	806
BIO	62,499	9197	108,434	69,230	13,710	776	9	0	1
WAT	62	53	71	33	16	28	0	1859	0
CON	299,358	186,107	24,1647	95,450	4927	3305	42	30,402	299,685
LTR	465,228	308,397	416,869	214,952	470,359	10,505	62,115	145,944	54,946
RLY	218,678	54,027	77,352	32,208	28,198	744	2773	8572	3718
AIR	33,234	21,421	15,620	7,991	2153	46	2312	697	1147
SEA	1017	1500	3189	362	32,522	25	57	869	543
HLM	0	0	0	0	0	0	0	0	0
SER	848,437	522,990	860,076	350,991	2,020,029	9655	56,472	179,594	192,200
Lab	3,689,555	2,645,224	13,081,859	4,985,874	7,939,400	15,5015	1,870,083	950,571	880,277
Cap	111,5629	799,304	3,842,846	1,408,910	7,823,192	16,6197	153,5925	2,387,623	2,272,012
Land	2,546,449	1,825,601	9,153,113	3,501,862					
RNASE									
RAL									
ROL									
RASE									
ROH									
USE									
USC									
UCL									
UOH									
PVT									
PUB									
GOV									
ITX	-1,139,709	-1,310,720	-1,108,597	-581,101	31,009	11,379	-202,612	90,587	81,514
CAC	53								
ROW	31,699	31,699	831,853	283,794	34,943	626,973	15,046	1,059,286	14,830,033
TOT	14,960,001	10,099,591	33,298,342	13,059,415	21,355,984	993915	4,030,348	5,696,101	19,224,020

	PAD	WHT	CER	CAS	ANH	FRS	FSH	COL	OIL
CO_2 (000' t)									
CH_4 (000' t)									
N_2O (000' t)									
Oil (mt)									34
coal (mt)								361	
ForestLand (Mha)									
Crop land (Mha)									
Grass land (Mha)									
Greenhouse effect (000' t)									

	GAS	FBV	TEX	WOD	MIN	PET	CHM	PAP	FER
PAD	0	480,042	41	8	3	0	20,196	1773	340
WHT	0	915,729	74	20	1	0	36,804	137	735
CER	1	3,051,005	7501	503	81	880	308,401	19,754	4427
CAS	2	5,426,983	2,056,049	341	201	1637	728,827	4,493	8404
ANH	3	1,601,763	372,582	325	153	308	157,594	882	2,331
FRS	4	19,532	290	55,073	37	217	16,076	50,715	69
FSH	0	458,661	40	2	1	0	19,972	172	356
COL	957	29,140	27,400	5,525	23,948	252,434	139,313	53,150	32,516
OIL	1376	194	12	300	58	18,673,034	211,974	556	1
GAS	5	3,010	36,448	183	1484	886	327,777	2027	381,218
FBV	2	4,300,382	16,609	498	218	1,015	479,422	11,745	10,967
TEX	19	106,573	5,501,608	1518	3789	4247	368,217	11,427	9356
WOD	2114	170,905	78,689	14,647	2562	9784	268,310	67,587	24,950
MIN	70	4328	6850	1884	26,313	2631	193,568	4602	162,315
PET	11,260	406,413	407,654	7406	63,514	1,020,947	1,182,041	84,364	791,718
CHM	28,662	1,241,295	2,158,888	38,087	121,069	546,658	15,269,445	421,793	1,275,236
PAP	325	442,279	160,128	19,331	1797	14,496	962,974	617,782	8541
FER	3	61,026	1978	6754	80	3612	225,431	525	577,350
CEM	53	76	370	140	218	3105	7641	70	55
IRS	329	1671	17,261	8151	12,877	4515	215,983	13,309	3104
ALU	116	3941	11,634	3771	49,536	10,981	225,889	7481	5077
OMN	19,255	49,225	196,681	18,950	64,147	20,287	564,638	12,863	9520
MCH	25,725	275,599	558,968	7848	33,712	41,103	484,615	15,348	23,692
NHY	15,172	218,487	616,474	13,510	61,600	294,141	789,216	97,953	79,410
HYD	2811	40,484	114,228	2503	11,414	54,502	146,236	18,150	14,714
NUC	427	6149	17,350	380	1734	8278	22,212	2757	2235
BIO	13	94,233	1534	185,645	130	781	59,890	172,836	313
WAT	133	6695	3641	46	404	209	11,973	115	1729

	GAS	FBV	TEX	WOD	MIN	PET	CHM	PAP	FER
CON	18,053	333,190	411,837	2372	58,265	192,032	328,759	9861	49,682
LTR	17,441	1,665,994	2,120,591	44,524	45,926	120,362	1,709,841	195,932	258,956
RLY	1848	93,918	29,913	4081	14,546	548,377	166,346	24,430	40,166
AIR	153	106,956	7532	3512	1299	15,347	31,377	5648	13,136
SEA	114	19,230	37,301	242	278	3792	52,835	5044	1304
HLM	0	0	0	0	0	0	0	0	0
SER	27,914	5,985,966	4,626,498	126,727	171,480	986,547	4,137,078	325,955	592,151
Lab	422,077	2,468,376	3,448,211	295,948	832,851	306,368	3,575,263	259,182	334,750
Cap	423,125	3,621,031	3,414,704	215,792	2,092,439	4,606,321	7,319,453	423,864	931,988
Land									
RNASE									
RAL									
ROL									
RASE									
ROH									
USE									
USC									
UCL									
UOH									
PVT									
PUB									
GOV									
ITX	13,229	768,166	483,529	20,759	61,959	1,554,145	2,787,084	233,237	301,979
CAC	CAC								
ROW	751,027	2,686,079	1,315,523	78,402	8,023,533	3,604,127	8,735,582	820,494	548,325
TOT	1,783,816	37,164,725	28,266,621	1,185,711	11,783,655	32,908,109	52,288,252	3,998,014	6,503,117

	GAS	FBV	TEX	WOD	MIN	PET	CHM	PAP	FER
CO_2 (000' t)									
CH_4 (000' t)									
N_2O (000' t)									
Oil (mt)									
coal (mt)									
ForestLand (Mha)									
Crop land (Mha)									
Grass land (Mha)									
Greenhouse effect (000' t)									

	CEM	IRS	ALU	OMN	MCH	NHY	HYD	NUC	BIO
PAD	0	19	1	1578	281	1309	0	0	5115
WHT	0	30	1	3020	546	2266	0	0	3042
CER	44	378	183	16,627	3852	10,779	0	0	21,3747
CAS	111	924	729	34,391	9627	16,922	0	0	2881
ANH	59	1459	1161	45,847	12,678	4697	0	0	6689
FRS	19	643	154	35,230	2603	704	0	0	1832
FSH	0	17	1	1639	343	1226	0	0	223
COL	217,084	2,065,513	457,147	574,505	1,140,56	1,692,816	0	0	164
OIL	2	20,936	1085	91,313	4544	40,917	0	0	209
GAS	35,930	317,837	26,419	70,721	35,564	325,683	0	0	27
FBV	166	2712	2049	22,239	6140	12,255	0	0	24,024
TEX	9504	11,516	4827	180,233	196,041	7383	0	0	3,233
WOD	28,946	16,799	5632	171,272	227,330	2250	0	0	350
MIN	368,646	651,084	477,031	1,175,277	213,464	0	0	6146	54
PET	126,628	734,155	145,332	863,213	521,420	1,654,974	0	0	28,271
CHM	137,084	308,761	232,203	2,018,159	2,645,906	117,745	0	1711	10,884
PAP	37,859	20,696	9488	186,071	248,553	27,143	7055	497	3004
FER	17	5835	198	6348	6177	3234	0	0	6528
CEM	1419	4361	1758	72,661	7136	31	24	0	9
IRS	3193	3,560,402	176,752	3,869,196	6,330,835	18,877	2020	0	147
ALU	1372	3,380,076	935,163	2,161,226	4,271,787	18,924	2551	0	212
OMN	171,932	1,036,459	217,809	7,001,479	4,244,945	361,252	12,119	5426	2,3057
MCH	7106	342,501	157,092	2,824,181	12,649,358	797,475	40,206	12,174	5690
NHY	248,364	1,113,522	190,742	874,539	603,835	4,222,060	832	61,371	1508
HYD	46,020	206,328	35,343	162,046	111,887	782,319	154	11,372	280
NUC	6990	31,339	5368	24,613	16,994	118,826	23	1727	42
BIO	176	2237	555	119,407	8788	2418	0	0	7717
WAT	23	633	187	20,829	1912	21,709	0	315	211
CON	9669	103,161	68,902	858,778	1,039,096	353,407	21,809	5453	25,876

	CEM	IRS	ALU	OMN	MCH	NHY	HYD	NUC	BIO
LTR	127,948	611,392	161,034	1,316,451	1293,680	389,306	1,1580	5827	118,111
RLY	147,032	955,566	161,873	524,171	21,1815	531,289	12,148	7897	8184
AIR	13,577	51,495	1,1825	41,876	12,219	53,022	406	776	635
SEA	655	3,886	1,032	32,578	33,574	4,699	853	81	2,670
HLM	0	0	0	0	0	0	0	0	0
SER	383,060	3,200,680	651,886	5,679,499	7,705,456	2,107,244	139,633	32,654	240,078
Lab	251,382	2,243,925	786,206	5,060,576	3,681,199	320,002	1,087,917	25,610	1,877,973
Cap	624,803	3,903,004	350,669	7,187,010	4,935,561	2,711,260	1,466,763	241,362	1,859,180
Land									
RNASE									
RAL									
ROL									
RASE									
ROH									
USE									
USC									
UCL									
UOH									
PVT									
PUB									
GOV									
ITX	111,024	1,029,909	356,922	2,600,023	4,055,593	−1,769,435	−32,813	834	0
CAC									
ROW	629,557	3,333,828	7,439,439	30,835,793	11,813,863	14,966,987	2,773,280	421,233	4,481,857
TOT	3,747,403	29,274,019	13,074,201	76,764,616	67,278,659				

	CEM	IRS	ALU	OMN	MCH	NHY	HYD	NUC	BIO
CO2 (000' t)									
CH4 (000' t)									
N2O (000' t)									
Oil (mt)									
coal (mt)									
ForestLand (Mha)									
Crop land (Mha)									
Grass land(Mha)									
Greenhouse effect (000' t)									

	WAT	CON	LTR	RLY	AIR	SEA	HLM	SER	Lab
PAD	49	113	18	0	0	111	3612	718,905	
WHT	93	80	1300	0	0	117	4530	397,700	
CER	388	1,061,416	953,328	0	0	712	13,580	2,341,984	
CAS	611	800	0	0	2	0	0	146,536	
ANH	178	601,652	0	0	0	0	10,338	1,285,048	
FRS	41	83,419	0	9	0	0	0	2992	
FSH	50	43	0	0	0	0	0	17,098	
COL	101	1970	0	4164	0	0	0	20,790	
OIL	619	23	0	0	11	0	0	13,295	
GAS	163	325	0	0	0	0	0	6777	
FBV	543	424	18,874	0	0	837	0	2,467,947	
TEX	112	93,123	79,766	910	50	941	17,421	152,856	
WOD	99	919,489	1068	97	0	0	0	23,867	
MIN	10	3,729,164	0	0	0	0	0	27,148	
PET	2190	3,355,210	12,200,344	268,248	97,767	43,643	70,748	950,541	
CHM	6332	1,141,838	2,324,087	8258	135,793	183,592	2,391,228	1,138,535	
PAP	1524	36,405	153,476	4697	951	460	14,464	287,397	
FER	1241	11,379	230	6	0	0	0	14,203	
CEM	0	3,969,131	0	0	0	0	0	1,738	
IRS	2159	10,273,278	891	409	0	0	0	207,244	
ALU	57	3853	326	0	0	0	0	122,373	
OMN	4979	8,386,495	2,113,947	1,250,215	91,698	103,333	133,745	2,538,691	
MCH	4336	2366,624	933,154	41763	5363	12,665	104,722	2,176,907	
NHY	22,460	895,618	28,247	695,572	2400	5157	15,150	647,088	
HYD	4162	165,952	5234	128,885	445	956	2807	119,901	
NUC	632	25,206	795	19,576	68	145	426	18,212	
BIO	140	285,604	4041	29	0	3	58	20,010	
WAT	141,549	177,244	15,485	320	985	18,334	900	109,167	

	WAT	CON	LTR	RLY	AIR	SEA	HLM	SER	Lab
CON	116,775	3,790,532	642,021	866,250	36,259	36,649	222,340	3,491,684	
LTR	6722	4,110,131	2,382,395	91,440	60,647	77,416	227,590	3,952,304	
RLY	656	981,304	532,260	568,506	1416	1228	1859	114,131	
AIR	47	92,231	142,044	3437	1177	512	1043	57,923	
SEA	183	25,850	41,715	1673	375	201	43,820	103,162	
HLM	0	0	0	100,676	0	0	0	173,246	
SER	89,571	9,607,585	8,210,058	232,405	112,249	176,900	761,775	19,477,882	
Lab	326,430	22,499,481	9,807,816	2,462,740	263,300	578,908	4,692,207	71,922,015	
Cap	363,570	9,461,837	7,531,709	1,774,196	197,957	415,166	2,693,040	95,799,645	
Land									
RNASE									13,686,272
RAL									30,655,386
ROL									9,597,032
RASE									23,604,200
ROH									6,014,659
USE									17,416,862
USC									63,190,125
UCL									9,330,416
UOH									2,282,323
PVT									
PUB									
GOV									
ITX	8260	2,468,171	2,627,536	224,044	44,498	48,704	414,210	790,126	
CAC	0	0							
ROW	0		478,851	0	0	0	0	7,888,498	
TOT	1,107,034	90,622,999	51,231,016	8,748,525	1,053,410	1,706,690	11,841,614	########	175,777,274

	WAT	CON	LTR	RLY	AIR	SEA	HLM	SER	LAB
CO2 (000't)									
CH4 (000't)									
N2O (000't)									
Oil (mt)									
coal (mt)									
ForestLand (Mha)									
Crop land (Mha)									
Grass land (Mha)									
Greenhouse effect (000't)									

	Cap	Land	RNASE	RAL	ROL	RASE	ROH	USE	USC
PAD			921,172	1,929,082	465,192	2,615,038	686,698	882,091	971,107
WHT			527,070	1,103,765	266,169	1,496,249	392,909	504,706	555,640
CER			1,827,586	3,590,841	947,921	4,944,834	1,532,174	2,474,139	2,817,419
CAS			299,626	627,464	151,311	850,582	223,359	286,913	315,868
ANH			1,203,885	1,541,555	566,288	4,038,074	1,205,631	2,093,656	2,433,331
FRS			56,650	106,912	28,156	168,129	46,044	38,522	42,409
FSH			270,564	566,605	136,635	768,082	201,695	259,085	285,231
COL			2587	4806	1392	7050	2297	3982	4539
OIL			0		0	0	0	0	0
GAS			8530	15,845	4591	23,244	7574	13,128	14,966
FBV			2,337,903	4,286,378	1,252,981	6,672,925	2,092,675	3,174,404	3,775,280
TEX			1,086,663	1,857,987	535,414	3,279,454	1,057,461	1,576,595	1,917,142
WOD			5587	5428	2970	18,109	5942	11,599	15,773
MIN			0	0	0	0	0	0	0
PET			345,403	571,748	238,310	1,112,680	650,630	723,282	1,928,089
CHM			419,787	578,193	220,158	1,431,805	465,963	850,403	1,564,865
PAP			27,994	36,969	13,736	96,565	26,651	56,321	93,756
FER			0	0	0	0	0	0	0
CEM			0	0	0	0	0	0	0
IRS			0	0	0	0	0	0	0
ALU			0	0	0	0	0	0	0
OMN			330,690	364,709	172,486	1,016,206	356,694	699,474	968,900
MCH			386,716	426,498	201,709	1,188,372	417,126	817,979	1,133,052
NHY			149,390	277,486	80,407	407,068	132,646	229,909	262,088
HYD			27,681	51,416	14,899	75,427	24,578	42,600	48,563
NUC			4204	7810	2263	11,457	3733	6471	7376
BIO			363,625	686,243	180,725	1,079,176	295,545	247,261	272,214
WAT			9931	18,447	5345	27,062	8818	15,284	17,423
CON			170,198	280,178	90,104	537,936	181,655	295,588	465,387

	Cap	Land	RNASE	RAL	ROL	RASE	ROH	USE	USC
LTR			161,6071	2,134,167	792,934	5,560,509	1,538,523	3,251,313	5,412,380
RLY			77,718	102,633	38,133	288,570	73,988	156,357	260,284
AIR			18,739	24,745	9194	−48,923	17,839	37,698	62,755
SEA			58,125	76,759	28,519	307,632	55,336	116,939	194,666
HLM			534,718	1,300,190	354,222	2,104,424	1,185,892	930,716	1,834,783
SER			4,613,267	6,644,682	2,597,189	15,922,245	6,145,036	10,965,155	21,129,082
Lab									
Cap									
Land		17,027,026							
RNASE	11,301,847								
RAL	99,289								
ROL	591,941								
RASE	28,570,049								
ROH	17,120,543								
USE	19,042,479								
USC	4,147,757								
UCL	1,346,862								
UOH	6,309,825								
PVT	32,937,007								
PUB	9,545,700								
GOV	7,439,300		355,411	0	0	4,142,314	1,419,397	0	2,379,634
ITX			802,814	1,317,718	426,876	2,548,941	867,247	1,406,262	2,238,702
CAC	44,737,987		10,554,871	5,666,113	2,167,887	18,770,185	5,912,228	10,750,880	25,845,192
ROW									
TOT	183,190,587	17,027,026	29,415,175	36,203,372	11,994,117	81,461,420	27,233,983	42,918,713	79,267,896

	CAP	LAND	RNASE	RAL	ROL	RASE	ROH	USE	USC
CO_2 (000' t)									
CH_4 (000' t)									
N_2O (000' t)									
Oil (mt)									
coal (mt)									
ForestLand (Mha)		0.16							
Crop land (Mha)		0.45							
Grass land (Mha)		0.01							
Greenhouse effect (000' t)	1,769,746								

	UCL	UOH	PVT	PUB	GOV	ITX	CAC	ROW	TOT
PAD	295,510	156,272			102,487		101,482	595,043	14,960,001
WHT	169,082	89,414			53,035		-115,959	533,924	10,099,591
CER	662,037	460,271			185,878		551,811	742,389	33,298,342
CAS	96,119	50,830			0		709,321	141,521	13,059,415
ANH	405,042	396,006			340,150		438,181	282,915	21,355,984
FRS	12,905	6825			61		4938	212,218	993,915
FSH	86,796	45,899			0		4929	656,848	4,030,348
COL	1797	815			6326		-80,482	16,566	5,696,101
OIL	0	0			0		-6482	112,740	19,224,020
GAS	5924	2688			27,381		11,623	75,561	1,783,816
FBV	892,075	647,357			398,033		1,400,206	2,500,625	37,164,725
TEX	342,432	326,420			328,519		594,522	8,394,477	28,266,621
WOD	1513	2298			107		-1,003,963	45,794	1,185,711
MIN	0	0			0		-130,444	4,862,982	11,783,655
PET	211,808	209,213			421,402		-2,900,636	2,787,765	32,908,109
CHM	188,901	230,639			908,608		4,992,972	6,114,416	52,288,252
PAP	11664	15059			80,533		94,443	167,706	3,998,014
FER	0	0			213		-58,531	240,268	6,503,117
CEM	0	0			0		-330,620	7135	3,747,403
IRS	0	0			0		1,842,627	2,699,301	29,274,019
ALU	0	0			0		543,543	1,313,233	13,074,201
OMN	100,834	150,272			539,350		21,869,319	21,128,635	76,764,616
MCH	117,917	175,731			580,572		30,745,595	6,497,608	67,278,659
NHY	103,738	47,068			529,604		0	0	14,966,987
HYD	19,222	8721			98,132		0	0	2,773,280
NUC	2920	1325			14,905		0	0	421,233
BIO	82,835	43,821			0		0	0	4,481,857
WAT	6896	3129			457,828		0	0	1,107,034
CON	69,489	84,079			738,347		73,456,402	0	90,622,999

	UCL	UOH	PVT	PUB	GOV	ITX	CAC	ROW	TOT
LTR	673,354	869,322			844,476		1,440,554	3,824,555	51,231,016
RLY	32,382	41,806			151,856		455,432	958,135	8,748,525
AIR	7807	10,080			20,100		33,413	106,138	1,053,410
SEA	24,218	31,267			45,962		197,583	112,451	1,706,690
HLM	322,525	810,765			2,189,457		0	0	11,841,614
SER	2,307,761	3,926,428			33,042,548		4,512,415	27,110,384	219,745,567
Lab								−251,300	175,777,274
Cap								−2,726,500	183,190,587
Land									17,027,026
RNASE					3,307,145			1,119,912	29,415,175
RAL					4,070,341			1,378,356	36,203,372
ROL					1,348,497			456,647	11,994,117
RASE					9,158,699			3,101,446	81,461,420
ROH					3,061,914			1,036,868	27,233,983
USE					4,825,346			1,634,026	42,918,713
USC					8,912,081			3,017,933	79,267,896
UCL					1,413,124			478,532	12,568,935
UOH					1,137,160			385,081	10,114,388
PVT					2,163,093				35,100,100
PUB				9,545,700					9,545,700
GOV	3,975,924	634,321	14,434,600				8,762,980	4,669,800	75,025,393
ITX	330,440,287	408,564	20,665,500			35,574,693		795,880	35,574,693
CAC	1,007,065	227,683			594,826			−641,415	148,137,174
ROW					−7,072,703				106,696,599
TOT	12,568,935	10,114,388	35100100	9545700	75,025,393	35,574,693	148,137,174	106,696,599	

	UCL	UOH	PVT	PUB	GOV	ITX	CAC	ROW	TOT
CO2 (000' t)									
CH4 (000' t)									
N2O (000' t)									
Oil (mt)									
coal (mt)									
ForestLand (Mha)									
Crop land (Mha)									
Grass land (Mha)									
Greenhouse effect (000' t)									

	CO2 (000' t)	CH4 (000' t)	N2O (000' t)	Oil (mt)	coal (mt)	ForestLand (Mha)	Crop land (Mha)	Grass land (Mha)	Greenhouse effect (000' t)
PAD	11,589	3327	41						94,028
WHT	8477	0	35						19,420
CER	18,240	0	40						30,705
CAS	3615	0	25						11,305
ANH	49,426	10,216	1						264,278
FRS	0	0	0						0
FSH	1	0	0						1
COL		730			8410				15,330
OIL		71		20					1495
GAS		708							14,872
FBV	27,626	7	0						27,837
TEX	1861	26	0						2413
WOD	351	0	0						353
MIN	1460	0	0						1465
PET	33,788	4	0						33,886
CHM	27,889	14	17						33,555
PAP	5223	25	0						5780
FER	31,514	3	1						31,765
CEM	129,920								129,920
IRS	116,958	15	1						117,621
ALU	2729	0	0						2729
OMN	41,822	1	1						42,092
MCH	17,662	0	0						17,776
NHY	715,830	980	11						739,704
HYD									0
NUC									0
BIO									0
WAT									0

	CO2 (000' t)	CH4 (000' t)	N2O (000' t)	Oil (mt)	coal (mt)	ForestLand (Mha)	Crop land (Mha)	Grass land (Mha)	Greenhouse effect (000' t)
CON	89	0	0						90
LTR	121,211	23	6						123,554
RLY	6109	0	2						10,211
AIR	10,122	0	0						1431
SEA	1416	0	0						0
HLM									
SER	1568	0	0						1583
Lab									0
Cap									0
Land						0.09	1.34	0.20	0
RNASE	16,306	190.189	3	21,215.69					21,216
RAL	29,861	323	5	38,226.28					38,226
ROL	8866	124	2	12,056.42					12,056
RASE	49,358	602	9	64,877.3					64,877
ROH	17,461	320	5	25,597.61					25,598
USE	8403	919	4	29,074.43					29,074
USC	22,105	1454	12	56,212.31					56,212
UCL	2492	284	1	8858.502					8858
UOH	2414	197	1	6943.122					6943
PVT									
PUB									
GOV									
ITX									
CAC									
ROW									
TOT									

	CO2 (000' t)	CH4 (000' t)	N2O (000' t)	Oil (mt)	coal (mt)	ForestLand (Mha)	Crop land (Mha)	Grass land (Mha)	Greenhouse effect (000' t)
CO2 (000' t)						67,800	207,520	38	−275,358
CH4 (000' t)									
N2O (000' t)									
Oil (mt)									
Coal (mt)									
ForestLand (Mha)									
Crop land (Mha)									
Grass land (Mha)	10,490								
Greenhouse effect (000' t)									

Appendix D: SAM Multiplier Matrix for India for the Year 2006–2007

	PAD	WHT	CER	CAS	ANH	FRS	FSH	COL	OIL
PAD	1.4828	0.1733	0.1871	0.1733	0.1777	0.0633	0.1667	0.1290	0.0358
WHT	0.1161	1.4727	0.1308	0.1135	0.1167	0.0414	0.1090	0.0844	0.0234
CER	0.3833	0.3861	1.4510	0.3934	0.5197	0.1433	0.3766	0.2938	0.0816
CAS	0.1355	0.1354	0.1423	1.2048	0.1433	0.0516	0.1397	0.1070	0.0295
ANH	0.2919	0.2484	0.2834	0.3042	1.2582	0.0924	0.2434	0.1898	0.0528
FRS	0.0089	0.0088	0.0093	0.0092	0.0092	1.0036	0.0090	0.0073	0.0020
FSH	0.0391	0.0386	0.0413	0.0409	0.0409	0.0149	1.1044	0.0303	0.0084
COL	0.0597	0.0613	0.0559	0.0556	0.0569	0.0211	0.0559	1.0508	0.0131
OIL	0.2242	0.2197	0.2014	0.2042	0.1948	0.0733	0.2109	0.1564	1.0472
GAS	0.0267	0.0299	0.0207	0.0223	0.0181	0.0065	0.0172	0.0148	0.0039
FBV	0.4021	0.3929	0.4190	0.4156	0.4336	0.1534	0.4065	0.3143	0.0872
TEX	0.2088	0.2075	0.2181	0.2162	0.2177	0.0803	0.2557	0.1641	0.0455
WOD	0.0116	0.0117	0.0111	0.0111	0.0110	0.0041	0.0126	0.0144	0.0026
MIN	0.0758	0.0763	0.0740	0.0744	0.0766	0.0284	0.0745	0.0611	0.0177
PET	0.3876	0.3795	0.3478	0.3527	0.3357	0.1264	0.3644	0.2693	0.0765
CHM	0.5026	0.5101	0.4892	0.4943	0.4920	0.1817	0.4806	0.4559	0.1106
PAP	0.0386	0.0384	0.0388	0.0388	0.0398	0.0148	0.0385	0.0333	0.0086
FER	0.2192	0.2695	0.1283	0.1558	0.0782	0.0258	0.0683	0.0532	0.0147
CEM	0.0434	0.0428	0.0436	0.0434	0.0460	0.0171	0.0445	0.0363	0.0108
IRS	0.2744	0.2719	0.2776	0.2761	0.2930	0.1090	0.2874	0.2369	0.0679
ALU	0.1097	0.1091	0.1112	0.1106	0.1171	0.0435	0.1148	0.0968	0.0272
OMN	0.6013	0.5930	0.6092	0.6064	0.6417	0.2407	0.6508	0.5271	0.1505
MCH	0.6683	0.6672	0.6803	0.6750	0.7153	0.2650	0.6932	0.6058	0.1682
NHY	0.1930	0.2083	0.1627	0.1600	0.1595	0.0586	0.1548	0.1516	0.0370
HYD	0.0358	0.0386	0.0302	0.0296	0.0296	0.0109	0.0287	0.0281	0.0069
NUC	0.0054	0.0059	0.0046	0.0045	0.0045	0.0016	0.0044	0.0043	0.0010
BIO	0.0573	0.0526	0.0578	0.0596	0.0547	0.0204	0.0520	0.0414	0.0113

	PAD	WHT	CER	CAS	ANH	FRS	FSH	COL	OIL
WAT	0.0111	0.0105	0.0118	0.0116	0.0126	0.0047	0.0117	0.0103	0.0028
CON	1.0583	1.0448	1.0649	1.0598	1.1251	0.4191	1.0886	0.8865	0.2607
LTR	0.5674	0.5627	0.5539	0.5557	0.5706	0.2111	0.5455	0.4448	0.1191
RLY	0.1052	0.0924	0.0865	0.0862	0.0872	0.0324	0.0842	0.0698	0.0192
AIR	0.0133	0.0133	0.0107	0.0109	0.0108	0.0040	0.0109	0.0084	0.0023
SEA	0.0182	0.0180	0.0190	0.0188	0.0205	0.0070	0.0182	0.0146	0.0040
HLM	0.1321	0.1279	0.1402	0.1390	0.1436	0.0532	0.1373	0.1091	0.0303
SER	2.1288	2.0666	2.1602	2.1496	2.3009	0.8242	2.1230	1.7382	0.4833
Lab	2.1773	2.1660	2.2805	2.2669	2.3397	0.8534	2.3166	1.6427	0.4569
Cap	2.0167	1.9859	2.0229	2.0197	2.3407	0.8784	2.2676	1.9423	0.5392
Land	0.4151	0.4381	0.4925	0.4812	0.2326	0.0715	0.1891	0.1467	0.0407
RNASE	0.3235	0.3169	0.3355	0.3338	0.3633	0.1348	0.3537	0.2776	0.0771
RAL	0.4172	0.4105	0.4396	0.4366	0.4545	0.1668	0.4463	0.3244	0.0902
ROL	0.1375	0.1352	0.1445	0.1436	0.1503	0.0552	0.1474	0.1082	0.0301
RASE	1.1039	1.1100	1.2059	1.1910	1.0135	0.3624	0.9463	0.7530	0.2091
ROH	0.2904	0.2835	0.2977	0.2965	0.3328	0.1244	0.3221	0.2654	0.0737
USE	0.4685	0.4586	0.4845	0.4822	0.5287	0.1966	0.5140	0.4083	0.1134
USC	0.9081	0.8930	0.9548	0.9486	0.9930	0.3649	0.9741	0.7151	0.1988
UCL	0.1430	0.1406	0.1501	0.1491	0.1571	0.0578	0.1539	0.1143	0.0318
UOH	0.1079	0.1054	0.1107	0.1102	0.1236	0.0462	0.1197	0.0985	0.0274
PVT	0.3819	0.3739	0.3854	0.3845	0.4449	0.1672	0.4296	0.3688	0.1024
PUB	0.1051	0.1035	0.1054	0.1052	0.1220	0.0458	0.1182	0.1012	0.0281
CAC	1.7102	1.6929	1.7605	1.7524	1.8830	0.6973	1.8262	1.4766	0.4101

	GAS	FBV	TEX	WOD	MIN	PET	CHM	PAP	FER
PAD	0.0966	0.1708	0.1474	0.1489	0.0511	0.0693	0.1181	0.1143	0.1186
WHT	0.0632	0.1381	0.0967	0.0976	0.0335	0.0453	0.0786	0.0747	0.0781
CER	0.2188	0.4468	0.3400	0.3460	0.1163	0.1580	0.2776	0.2688	0.2732
CAS	0.0791	0.2993	0.2186	0.1225	0.0422	0.0573	0.1199	0.0979	0.1039
ANH	0.1411	0.2807	0.2373	0.2178	0.0752	0.1023	0.1782	0.1677	0.1757
FRS	0.0052	0.0091	0.0083	0.0553	0.0028	0.0038	0.0075	0.0222	0.0068
FSH	0.0228	0.0500	0.0346	0.0349	0.0120	0.0162	0.0281	0.0267	0.0279
COL	0.0347	0.0531	0.0572	0.0578	0.0210	0.0344	0.0494	0.0611	0.0521
OIL	0.1142	0.1898	0.1968	0.1789	0.0619	0.6704	0.1676	0.1589	0.2284
GAS	1.0104	0.0179	0.0193	0.0167	0.0058	0.0080	0.0226	0.0148	0.0795
FBV	0.2343	1.4948	0.3596	0.3620	0.1245	0.1690	0.2989	0.2810	0.2927
TEX	0.1220	0.1929	1.4276	0.1889	0.0653	0.0882	0.1599	0.1483	0.1537
WOD	0.0077	0.0158	0.0144	1.0228	0.0037	0.0053	0.0160	0.0289	0.0142
MIN	0.0441	0.0676	0.0682	0.0687	1.0266	0.0335	0.0598	0.0538	0.0836
PET	0.1954	0.3274	0.3390	0.3078	0.1066	1.1752	0.2797	0.2732	0.3944
CHM	0.2993	0.4923	0.5680	0.4776	0.1639	0.2322	1.7659	0.5132	0.6600
PAP	0.0231	0.0523	0.0464	0.0553	0.0125	0.0177	0.0593	1.2143	0.0370
FER	0.0395	0.0913	0.0692	0.0681	0.0210	0.0287	0.0576	0.0488	1.1482
CEM	0.0264	0.0400	0.0403	0.0398	0.0145	0.0203	0.0324	0.0309	0.0330
IRS	0.1698	0.2554	0.2602	0.2632	0.0939	0.1278	0.2137	0.2031	0.2122
ALU	0.0685	0.1028	0.1060	0.1074	0.0418	0.0514	0.0911	0.0836	0.0871
OMN	0.3765	0.5600	0.5715	0.5756	0.2057	0.2815	0.4663	0.4412	0.4652
MCH	0.4225	0.6274	0.6490	0.6266	0.2258	0.3114	0.5133	0.4881	0.5162
NHY	0.1025	0.1548	0.1817	0.1572	0.0566	0.0841	0.1454	0.1549	0.1433
HYD	0.0190	0.0287	0.0337	0.0291	0.0105	0.0156	0.0269	0.0287	0.0266
NUC	0.0029	0.0044	0.0051	0.0044	0.0016	0.0024	0.0041	0.0044	0.0040
BIO	0.0303	0.0525	0.0476	0.2059	0.0161	0.0218	0.0408	0.0901	0.0385
WAT	0.0073	0.0113	0.0114	0.0111	0.0040	0.0057	0.0098	0.0092	0.0099
CON	0.6451	0.9764	0.9817	0.9697	0.3550	0.4913	0.7846	0.7521	0.8042

	GAS	FBV	TEX	WOD	MIN	PET	CHM	PAP	FER
LTR	0.3176	0.5397	0.5768	0.5193	0.1689	0.2315	0.4295	0.4346	0.4402
RLY	0.0503	0.0808	0.0800	0.0801	0.0281	0.0554	0.0676	0.0694	0.0734
AIR	0.0061	0.0130	0.0101	0.0125	0.0034	0.0050	0.0086	0.0093	0.0102
SEA	0.0107	0.0172	0.0181	0.0167	0.0057	0.0079	0.0145	0.0143	0.0137
HLM	0.0808	0.1237	0.1227	0.1240	0.0432	0.0600	0.0996	0.0965	0.1009
SER	1.2617	2.1339	2.1555	2.0445	0.6857	0.9734	1.6868	1.6358	1.7174
Lab	1.3085	1.9601	1.9447	2.0184	0.6513	0.8494	1.4954	1.4606	1.5013
Cap	1.3359	1.9730	1.9987	2.0054	0.7747	1.0542	1.6486	1.5797	1.7018
Land	0.1092	0.2571	0.1947	0.1710	0.0580	0.0788	0.1428	0.1331	0.1373
RNASE	0.2055	0.3065	0.3075	0.3136	0.1103	0.1491	0.2471	0.2393	0.2513
RAL	0.2551	0.3825	0.3806	0.3933	0.1285	0.1708	0.2974	0.2901	0.2989
ROL	0.0844	0.1265	0.1260	0.1300	0.0429	0.0571	0.0988	0.0963	0.0994
RASE	0.5520	0.9170	0.8583	0.8453	0.2989	0.4070	0.6809	0.6534	0.6856
ROH	0.1893	0.2812	0.2837	0.2868	0.1056	0.1442	0.2321	0.2236	0.2376
USE	0.2995	0.4462	0.4483	0.4562	0.1622	0.2200	0.3618	0.3499	0.3685
USC	0.5578	0.8359	0.8327	0.8591	0.2834	0.3776	0.6530	0.6365	0.6574
UCL	0.0883	0.1323	0.1319	0.1359	0.0453	0.0605	0.1039	0.1011	0.1048
UOH	0.0703	0.1045	0.1054	0.1065	0.0392	0.0535	0.0862	0.0830	0.0882
PVT	0.2541	0.3758	0.3808	0.3820	0.1470	0.2013	0.3154	0.3024	0.3252
PUB	0.0696	0.1028	0.1041	0.1045	0.0404	0.0549	0.0859	0.0823	0.0887
CAC	1.0628	1.5995	1.5951	1.6115	0.5877	0.7924	1.2864	1.2390	1.3144

	CEM	IRS	ALU	OMN	MCH	NHY	HYD	NUC	BIO
PAD	0.1115	0.1144	0.0535	0.0790	0.1052	0.1300	0.1712	0.1549	0.1744
WHT	0.0730	0.0749	0.0350	0.0518	0.0689	0.0853	0.1120	0.1013	0.1146
CER	0.2549	0.2606	0.1218	0.1805	0.2403	0.2989	0.3875	0.3535	0.4409
CAS	0.0927	0.0941	0.0442	0.0660	0.0876	0.1094	0.1397	0.1280	0.1419
ANH	0.1641	0.1680	0.0784	0.1167	0.1549	0.1920	0.2504	0.2289	0.2545
FRS	0.0066	0.0063	0.0030	0.0050	0.0061	0.0073	0.0092	0.0085	0.0096
FSH	0.0261	0.0268	0.0125	0.0185	0.0246	0.0306	0.0403	0.0364	0.0407
COL	0.1103	0.1345	0.0610	0.0452	0.0561	0.2098	0.0577	0.0795	0.0571
OIL	0.1620	0.1589	0.0731	0.1055	0.1362	0.2539	0.1894	0.1933	0.1975
GAS	0.0242	0.0266	0.0088	0.0111	0.0146	0.0452	0.0177	0.0213	0.0179
FBV	0.2719	0.2784	0.1302	0.1926	0.2564	0.3179	0.4154	0.3779	0.4232
TEX	0.1445	0.1450	0.0680	0.1031	0.1374	0.1653	0.2160	0.1968	0.2176
WOD	0.0164	0.0093	0.0046	0.0085	0.0123	0.0110	0.0110	0.0108	0.0111
MIN	0.1519	0.0845	0.0648	0.0570	0.0609	0.0664	0.0780	0.0906	0.0769
PET	0.2793	0.2728	0.1258	0.1795	0.2342	0.4339	0.3266	0.3326	0.3407
CHM	0.3929	0.3659	0.1880	0.2783	0.3855	0.4251	0.4889	0.4715	0.4928
PAP	0.0406	0.0293	0.0145	0.0233	0.0327	0.0360	0.0424	0.0394	0.0403
FER	0.0462	0.0472	0.0221	0.0328	0.0438	0.0541	0.0697	0.0635	0.0750
CEM	1.0317	0.0319	0.0148	0.0231	0.0295	0.0393	0.0470	0.0456	0.0462
IRS	0.2045	1.3480	0.1126	0.2087	0.3238	0.2601	0.3012	0.2942	0.2943
ALU	0.0838	0.2268	1.1188	0.1013	0.1785	0.1087	0.1214	0.1192	0.1175
OMN	0.4958	0.4980	0.2272	1.4115	0.4957	0.5811	0.6553	0.6481	0.6478
MCH	0.4942	0.5155	0.2432	0.3869	1.6826	0.6820	0.7415	0.7414	0.7167
NHY	0.2108	0.1824	0.0782	0.1012	0.1306	1.5374	0.1596	0.3588	0.1593
HYD	0.0391	0.0338	0.0145	0.0188	0.0242	0.0996	1.0296	0.0665	0.0295
NUC	0.0059	0.0051	0.0022	0.0028	0.0037	0.0151	0.0045	1.0101	0.0045
BIO	0.0369	0.0360	0.0169	0.0270	0.0341	0.0416	0.0533	0.0487	1.0553
WAT	0.0089	0.0091	0.0043	0.0067	0.0087	0.0116	0.0126	0.0132	0.0126
CON	0.7626	0.7740	0.3559	0.5365	0.7117	0.9555	1.1503	1.1143	1.1312

	CEM	IRS	ALU	OMN	MCH	NHY	HYD	NUC	BIO
LTR	0.4045	0.4011	0.1891	0.2794	0.3730	0.4712	0.5473	0.5244	0.5718
RLY	0.1056	0.1046	0.0439	0.0532	0.0665	0.1274	0.0903	0.1101	0.0877
AIR	0.0113	0.0097	0.0045	0.0059	0.0075	0.0137	0.0107	0.0125	0.0108
SEA	0.0127	0.0129	0.0060	0.0093	0.0124	0.0151	0.0191	0.0177	0.0194
HLM	0.0946	0.0971	0.0454	0.0671	0.0898	0.1059	0.1430	0.1311	0.1433
SER	1.6090	1.6661	0.7739	1.1533	1.5874	1.8995	2.2432	2.1573	2.2515
Lab	1.4286	1.4787	0.7093	1.0235	1.3546	1.6785	2.2845	1.8930	2.3382
Cap	1.6281	1.6313	0.7198	1.1043	1.4496	2.0340	2.4571	2.5003	2.3581
Land	0.1271	0.1299	0.0608	0.0901	0.1199	0.1490	0.1933	0.1762	0.2096
RNASE	0.2378	0.2429	0.1125	0.1670	0.2215	0.2796	0.3664	0.3369	0.3644
RAL	0.2821	0.2921	0.1400	0.2027	0.2698	0.3227	0.4452	0.3749	0.4544
ROL	0.0939	0.0971	0.0463	0.0673	0.0895	0.1078	0.1477	0.1258	0.1503
RASE	0.6451	0.6579	0.3040	0.4530	0.6016	0.7565	0.9857	0.9179	0.9934
ROH	0.2252	0.2281	0.1035	0.1560	0.2065	0.2692	0.3420	0.3311	0.3345
USE	0.3488	0.3556	0.1639	0.2442	0.3237	0.4119	0.5357	0.4989	0.5306
USC	0.6207	0.6415	0.3060	0.4447	0.5915	0.7126	0.9765	0.8321	0.9933
UCL	0.0989	0.1021	0.0485	0.0706	0.0939	0.1141	0.1551	0.1339	0.1572
UOH	0.0836	0.0847	0.0384	0.0579	0.0767	0.0999	0.1270	0.1228	0.1243
PVT	0.3098	0.3110	0.1378	0.2111	0.2780	0.3810	0.4660	0.4726	0.4481
PUB	0.0848	0.0850	0.0375	0.0575	0.0755	0.1060	0.1280	0.1303	0.1229
CAC	1.2526	1.2684	0.5756	0.8656	1.1408	1.5269	1.9205	1.8392	1.8865

	WAT	CON	LTR	RLY	AIR	SEA	HLM	SER	Lab
PAD	0.1672	0.1421	0.1376	0.1497	0.1460	0.1581	0.1622	0.1674	0.1880
WHT	0.1095	0.0932	0.0903	0.0980	0.0958	0.1037	0.1067	0.1096	0.1233
CER	0.3806	0.3364	0.3326	0.3404	0.3334	0.3601	0.3691	0.3821	0.4206
CAS	0.1372	0.1161	0.1133	0.1223	0.1220	0.1313	0.1362	0.1356	0.1508
ANH	0.2453	0.2145	0.2018	0.2193	0.2142	0.2313	0.2376	0.2451	0.2698
FRS	0.0092	0.0091	0.0075	0.0083	0.0080	0.0086	0.0088	0.0087	0.0098
FSH	0.0394	0.0333	0.0322	0.0352	0.0343	0.0372	0.0381	0.0385	0.0445
COL	0.0620	0.0663	0.0502	0.0691	0.0533	0.0554	0.0556	0.0557	0.0570
OIL	0.1944	0.1961	0.3038	0.2016	0.2323	0.2016	0.1919	0.1864	0.1998
GAS	0.0186	0.0177	0.0151	0.0188	0.0168	0.0176	0.0186	0.0171	0.0181
FBV	0.4060	0.3443	0.3343	0.3629	0.3557	0.3849	0.3940	0.4076	0.4519
TEX	0.2107	0.1802	0.1751	0.1893	0.1854	0.2006	0.2072	0.2060	0.2341
WOD	0.0123	0.0211	0.0098	0.0115	0.0112	0.0114	0.0122	0.0108	0.0113
MIN	0.0816	0.1165	0.0634	0.0757	0.0693	0.0729	0.0732	0.0747	0.0754
PET	0.3340	0.3385	0.5280	0.3476	0.4011	0.3471	0.3292	0.3213	0.3448
CHM	0.4900	0.4400	0.4704	0.4453	0.6105	0.6075	0.7460	0.4747	0.5026
PAP	0.0410	0.0357	0.0377	0.0367	0.0397	0.0406	0.0449	0.0395	0.0408
FER	0.0698	0.0592	0.0577	0.0613	0.0609	0.0655	0.0679	0.0683	0.0760
CEM	0.0506	0.0835	0.0375	0.0451	0.0407	0.0430	0.0434	0.0449	0.0452
IRS	0.3082	0.3866	0.2425	0.2831	0.2614	0.2768	0.2772	0.2866	0.2889
ALU	0.1189	0.1213	0.0986	0.1131	0.1060	0.1120	0.1120	0.1149	0.1159
OMN	0.6489	0.6491	0.5681	0.7452	0.6501	0.6596	0.6132	0.6300	0.6370
MCH	0.7119	0.6282	0.5987	0.6480	0.6205	0.6675	0.6765	0.6986	0.7098
NHY	0.1917	0.1656	0.1378	0.2667	0.1482	0.1566	0.1577	0.1564	0.1623
HYD	0.0355	0.0307	0.0255	0.0494	0.0275	0.0290	0.0292	0.0290	0.0301
NUC	0.0054	0.0047	0.0039	0.0075	0.0042	0.0044	0.0044	0.0044	0.0046
BIO	0.0527	0.0494	0.0431	0.0474	0.0461	0.0495	0.0510	0.0509	0.0575
WAT	1.1593	0.0131	0.0111	0.0117	0.0124	0.0243	0.0123	0.0127	0.0125
CON	1.2275	1.9629	0.9131	1.0904	0.9881	1.0486	1.0574	1.0967	1.1044

	WAT	CON	LTR	RLY	AIR	SEA	HLM	SER	Lab
LTR	0.5450	0.5105	1.4923	0.4997	0.5348	0.5554	0.5414	0.5374	0.5745
RLY	0.0881	0.0938	0.0862	1.1539	0.0795	0.0821	0.0820	0.0828	0.0868
AIR	0.0107	0.0107	0.0117	0.0105	1.0106	0.0104	0.0103	0.0104	0.0110
SEA	0.0186	0.0160	0.0161	0.0168	0.0167	1.0176	0.0216	0.0184	0.0199
HLM	0.1396	0.1189	0.1160	0.1376	0.1229	0.1322	1.1353	0.1367	0.1530
SER	2.2632	1.9976	1.9880	2.0081	2.0474	2.1702	2.1717	3.1887	2.2899
Lab	2.2478	1.9343	1.8149	2.0382	1.9544	2.1478	2.2232	2.1626	2.9423
Cap	2.3271	1.8704	1.8667	2.0231	1.9730	2.1174	2.1197	2.3089	1.9735
Land	0.1897	0.1646	0.1615	0.1696	0.1665	0.1798	0.1849	0.1897	0.2103
RNASE	0.3551	0.2980	0.2890	0.3167	0.3079	0.3333	0.3402	0.3465	0.3868
RAL	0.4383	0.3777	0.3576	0.3974	0.3838	0.4194	0.4336	0.4223	0.5584
ROL	0.1451	0.1247	0.1184	0.1314	0.1270	0.1386	0.1430	0.1401	0.1817
RASE	0.9557	0.8046	0.7866	0.8507	0.8309	0.8967	0.9146	0.9388	1.0127
ROH	0.3282	0.2706	0.2667	0.2896	0.2828	0.3042	0.3078	0.3228	0.3184
USE	0.5180	0.4327	0.4214	0.4607	0.4484	0.4847	0.4936	0.5063	0.5491
USC	0.9593	0.8238	0.7825	0.8680	0.8389	0.9156	0.9451	0.9257	1.1992
UCL	0.1520	0.1301	0.1240	0.1373	0.1328	0.1447	0.1491	0.1470	0.1860
UOH	0.1219	0.1005	0.0991	0.1076	0.1050	0.1130	0.1144	0.1199	0.1185
PVT	0.4423	0.3572	0.3569	0.3855	0.3770	0.4039	0.4049	0.4384	0.3783
PUB	0.1213	0.0975	0.0973	0.1054	0.1028	0.1103	0.1105	0.1203	0.1028
CAC	1.8429	1.5186	1.4867	1.6258	1.5790	1.7064	1.7285	1.8114	1.8383

	Cap	Land	RNASE	RAL	ROL	RASE	ROH	USE	USC
PAD	0.1633	0.1962	0.1909	0.2273	0.2057	0.1962	0.1868	0.1784	0.1647
WHT	0.1065	0.1285	0.1248	0.1486	0.1349	0.1285	0.1226	0.1174	0.1081
CER	0.3724	0.4276	0.4191	0.4752	0.4487	0.4276	0.4207	0.4182	0.3860
CAS	0.1352	0.1551	0.1515	0.1710	0.1615	0.1551	0.1514	0.1476	0.1378
ANH	0.2425	0.2825	0.2681	0.2827	0.2821	0.2825	0.2759	0.2769	0.2535
FRS	0.0089	0.0104	0.0101	0.0114	0.0107	0.0104	0.0100	0.0091	0.0087
FSH	0.0384	0.0465	0.0453	0.0542	0.0490	0.0465	0.0442	0.0422	0.0388
COL	0.0582	0.0570	0.0576	0.0572	0.0568	0.0570	0.0567	0.0569	0.0568
OIL	0.1836	0.1987	0.1924	0.2028	0.2029	0.1987	0.2021	0.1986	0.1989
GAS	0.0175	0.0183	0.0182	0.0191	0.0186	0.0183	0.0181	0.0180	0.0175
FBV	0.3992	0.4621	0.4499	0.5069	0.4869	0.4621	0.4558	0.4477	0.4141
TEX	0.2086	0.2419	0.2333	0.2579	0.2470	0.2419	0.2396	0.2343	0.2166
WOD	0.0106	0.0115	0.0112	0.0117	0.0117	0.0115	0.0114	0.0114	0.0110
MIN	0.0806	0.0752	0.0775	0.0740	0.0741	0.0752	0.0748	0.0753	0.0765
PET	0.3164	0.3428	0.3318	0.3500	0.3502	0.3428	0.3487	0.3426	0.3431
CHM	0.4869	0.5047	0.4951	0.5080	0.5049	0.5047	0.5058	0.5014	0.4984
PAP	0.0389	0.0411	0.0402	0.0416	0.0413	0.0411	0.0407	0.0408	0.0400
FER	0.0668	0.0785	0.0765	0.0888	0.0821	0.0785	0.0759	0.0737	0.0682
CEM	0.0489	0.0450	0.0466	0.0441	0.0441	0.0450	0.0447	0.0451	0.0460
IRS	0.3101	0.2879	0.2969	0.2816	0.2831	0.2879	0.2863	0.2890	0.2940
ALU	0.1230	0.1156	0.1186	0.1132	0.1140	0.1156	0.1151	0.1163	0.1177
OMN	0.6707	0.6360	0.6500	0.6220	0.6277	0.6360	0.6329	0.6404	0.6461
MCH	0.7512	0.7084	0.7262	0.6925	0.6992	0.7084	0.7052	0.7137	0.7204
NHY	0.1579	0.1632	0.1621	0.1688	0.1653	0.1632	0.1620	0.1614	0.1573
HYD	0.0293	0.0302	0.0300	0.0313	0.0306	0.0302	0.0300	0.0299	0.0291
NUC	0.0044	0.0046	0.0046	0.0048	0.0047	0.0046	0.0046	0.0045	0.0044
BIO	0.0513	0.0617	0.0598	0.0685	0.0638	0.0617	0.0591	0.0534	0.0503
WAT	0.0130	0.0126	0.0123	0.0124	0.0123	0.0126	0.0126	0.0123	0.0123
CON	1.1990	1.0998	1.1402	1.0763	1.0785	1.0998	1.0927	1.1016	1.1265

	Cap	Land	RNASE	RAL	ROL	RASE	ROH	USE	USC
LTR	0.5303	0.5815	0.5591	0.5791	0.5795	0.5815	0.5674	0.5820	0.5678
RLY	0.0858	0.0876	0.0867	0.0873	0.0869	0.0876	0.0861	0.0869	0.0863
AIR	0.0103	0.0099	0.0111	0.0115	0.0114	0.0099	0.0111	0.0113	0.0110
SEA	0.0185	0.0212	0.0191	0.0197	0.0198	0.0212	0.0195	0.0199	0.0195
HLM	0.1402	0.1529	0.1409	0.1628	0.1550	0.1529	0.1706	0.1456	0.1466
SER	2.1694	2.2769	2.1852	2.2604	2.2756	2.2769	2.3025	2.2962	2.2986
Lab	1.8582	1.9557	1.9119	1.9962	1.9680	1.9557	1.9518	1.9357	1.9019
Cap	2.8874	1.9804	1.9276	1.9976	1.9868	1.9804	1.9841	1.9725	1.9502
Land	0.1857	1.2157	0.2109	0.2420	0.2260	0.2157	0.2102	0.2061	0.1906
RNASE	0.3623	0.3110	1.3021	0.3129	0.3100	0.3110	0.3110	0.3067	0.3039
RAL	0.3743	0.3871	0.3767	1.3914	0.3864	0.3871	0.3866	0.3808	0.3764
ROL	0.1269	0.1281	0.1246	0.1294	1.1278	0.1281	0.1279	0.1260	0.1246
RASE	0.9950	1.8884	0.8633	0.9164	0.8949	1.8884	0.8832	0.8686	0.8484
ROH	0.3700	0.2858	0.2773	0.2867	0.2847	0.2858	1.2861	0.2823	0.2802
USE	0.5419	0.4530	0.4399	0.4554	0.4515	0.4530	0.4531	1.4468	0.4429
USC	0.8399	0.8464	0.8234	0.8551	0.8447	0.8464	0.8453	0.8329	1.8234
UCL	0.1368	0.1340	0.1303	0.1353	0.1337	0.1340	0.1339	0.1319	0.1304
UOH	0.1372	0.1062	0.1030	0.1065	0.1058	0.1062	0.1063	0.1049	0.1041
PVT	0.5450	0.3800	0.3690	0.3816	0.3796	0.3800	0.3807	0.3771	0.3738
PUB	0.1505	0.1032	0.1004	0.1041	0.1035	0.1032	0.1034	0.1028	0.1016
CAC	2.0367	1.8275	1.9143	1.7763	1.7842	1.8275	1.8139	1.8340	1.8872

	UCL	UOH	PVT	PUB	CAC
PAD	0.1945	0.1799	0.1501	0.1298	0.1298
WHT	0.1278	0.1187	0.0977	0.0837	0.0837
CER	0.4353	0.4205	0.3457	0.3049	0.3049
CAS	0.1553	0.1490	0.1261	0.1134	0.1134
ANH	0.2771	0.2772	0.2236	0.1960	0.1960
FRS	0.0097	0.0092	0.0083	0.0076	0.0076
FSH	0.0459	0.0424	0.0349	0.0303	0.0303
COL	0.0572	0.0557	0.0588	0.0601	0.0601
OIL	0.2057	0.2082	0.1762	0.1603	0.1603
GAS	0.0187	0.0180	0.0173	0.0166	0.0166
FBV	0.4724	0.4557	0.3691	0.3233	0.3233
TEX	0.2373	0.2377	0.1926	0.1694	0.1694
WOD	0.0116	0.0117	0.0102	0.0094	0.0094
MIN	0.0730	0.0717	0.0826	0.0891	0.0891
PET	0.3549	0.3594	0.3034	0.2756	0.2756
CHM	0.5114	0.5256	0.4805	0.4592	0.4592
PAP	0.0418	0.0423	0.0382	0.0356	0.0356
FER	0.0787	0.0745	0.0618	0.0541	0.0541
CEM	0.0435	0.0426	0.0504	0.0550	0.0550
IRS	0.2789	0.2743	0.3185	0.3449	0.3449
ALU	0.1124	0.1112	0.1257	0.1345	0.1345
OMN	0.6197	0.6156	0.6835	0.7246	0.7246
MCH	0.6879	0.6817	0.7663	0.8181	0.8181
NHY	0.1715	0.1622	0.1577	0.1495	0.1495
HYD	0.0318	0.0300	0.0292	0.0277	0.0277
NUC	0.0048	0.0046	0.0044	0.0042	0.0042
BIO	0.0573	0.0539	0.0473	0.0423	0.0423
WAT	0.0150	0.0130	0.0149	0.0115	0.0115
CON	1.0620	1.0383	1.2372	1.3551	1.3551

	UCL	UOH	PVT	PUB	CAC
LTR	0.5858	0.6090	0.5060	0.4634	0.4634
RLY	0.0873	0.0871	0.0853	0.0842	0.0842
AIR	0.0114	0.0116	0.0101	0.0096	0.0096
SEA	0.0202	0.0211	0.0176	0.0160	0.0160
HLM	0.1677	0.2126	0.1364	0.1095	0.1095
SER	2.4502	2.5300	2.2046	1.8716	1.8716
Lab	2.0073	2.0295	1.8407	1.6951	1.6951
Cap	2.0551	2.0889	1.8817	1.7056	1.7056
Land	0.2175	0.2076	0.1720	0.1515	0.1515
RNASE	0.3321	0.3252	0.3114	0.2693	0.2693
RAL	0.4116	0.4023	0.3860	0.3360	0.3360
ROL	0.1362	0.1332	0.1278	0.1111	0.1111
RASE	0.9434	0.9121	0.8566	0.7339	0.7339
ROH	0.3062	0.3002	0.2870	0.2471	0.2471
USE	0.4841	0.4742	0.4538	0.3920	0.3920
USC	0.9003	0.8802	0.8444	0.7344	0.7344
UCL	1.1426	0.1395	0.1337	0.1162	0.1162
UOH	0.1137	1.1115	0.1066	0.0918	0.0918
PVT	0.4016	0.4006	1.3723	0.3276	0.3276
PUB	0.1071	0.1088	0.0981	1.0889	0.0889
CAC	1.7415	1.6939	2.1114	2.3708	2.3708

Appendix E: The description of 60 sectors of 1994–1995 IO Tables

Code	Sectoral description	Code	Sectoral description
S1	Food crops	S31	Paints, varnishes and lacquers
S2	Cash crops	S32	Pesticides, drugs and other chemicals
S3	Plantation crops	S33	Cement
S4	Other crops	S34	Non-metallic mineral products
S5	Animal husbandry	S35	Iron and steel industries and foundries
S6	Forestry and logging	S36	Other basic metal industry
S7	Fishing	S37	Metal products except machinery
S8	Coal and lignite	S38	Agricultural machinery
S9	Crude petroleum and natural gas	S39	Machinery for food and textiles
S10	Irone ore	S40	Other machinery
S11	Other minerals	S41	Electrical, electronic, machinery and appliances
S12	Sugar	S42	Railway transport equipment
S13	Food products excluding sugar	S43	Other transport equipment
S14	Beverages	S44	Miscellaneous manufacturing industries
S15	Tobacco products	S45	Construction
S16	Cotton textiles	S46	Electricity
S17	Wool, silk, and synthetic fibre textiles	S47	Gas and water supply
S18	Jute, hemp and mesta textiles	S48	Railway transport services
S19	Textiles products including wearing apparel	S49	Other transport services
S20	Wood and wood products except furniture	S50	Storage and warehousing
S21	Furniture and fixture	S51	Communication
S22	Paper and paper products	S52	Trade
S23	Printing, publishing and allied activities	S53	Hotels and restaurants
S24	Leather and leather products	S54	Banking
S25	Plastic and rubber products	S55	Insurance
S26	Petroleum products	S56	Ownership of dwellings
S27	Coal tar products	S57	Education and research
S28	Inorganic heavy chemicals	S58	Medical and health
S29	Organic heavy chemicals	S59	Other services
S30	Fertilisers	S60	Public administration and defence

Appendix F: 35 Sectors Input–Output Tables of 1994–1995 (₹ Lakhs)

	PAD	WHT	CER	CAS	ANH	FRS	FSH	COL
PAD	631,133	2739	125,280	0	104,977	0	0	0
WHT	1971	328,910	70,051	0	55,446	0	0	0
CER	21,360	18,036	512,690	0	1,747,879	552	570	10
CAS	542	32	27,750	190,556	14,537	0	0	0
ANH	335,015	119,944	470,532	183,602	5010	0	36	0
FRS	41	48	685	0	3	0	0	0
FSH	28	12	1032	0	36	0	0	0
COL	804	235	2616	0	3	3515	15,660	12,722
OIL	0	0	0	0	1	0	0	0
GAS	2	1	58	2	0	1	0	26
FBV	583	161	4119	0	178,539	16	1	0
TXL	5030	2091	3296	0	46,054	958	4010	154
WOD	204	133	320	0	3	148	28,424	9852
MIN	0	0	68	0	0	0	5805	9139
PET	64,990	34,895	74,897	15,081	12	4687	0	35,500
CHM	30,944	20,655	40,316	59,351	8904	1483	27,586	35,740
PAP	237	152	967	43	17	712	1661	1850
FER	418,715	267,292	504,434	227,614	586	16	123	2302
CEM	0	0	0	0	0	3	89	0
IRS	0	0	8	0	0	174	0	0
ALU	0	0	104	0	0	24	461	0
OMN	7020	4369	9904	2737	3520	4429	64	34,164
MCH	41,437	27,557	70,451	15,862	2	646	28,579	134,568
HYD	13,286	20,950	16,566	3634	2	119	222	20,584
NHY	42,328	66,743	52,778	11,577	6	380	294	65,579
NUC	899	1417	1120	246	0	8	6	1392

	PAD	WHT	CER	CAS	ANH	FRS	FSH	COL
BIO	29,528	10,649	45,088	16,093	14,816	4	38	0
WAT	168	135	205	47	1	25	32	573
CON	110,415	78,169	135,296	44,665	8862	7065	0	2240
LTR	25,483	12,669	39,495	14,477	42,460	4280	2954	16,277
RLY	28,696	19,418	29,715	10,601	4155	1534	1466	9778
AIR	1616	803	2505	918	2692	271	187	1032
SEA	3394	1687	5260	1928	5654	570	393	2168
HLM	0	0	0	0	0	2892	0	0
SER	285,696	121,770	256,678	116,539	363,864	15,809	23,832	103,075
GVA	4,100,818	2,160,915	7,734,042	4,051,046	5,264,696	487,069	1,030,604	643,868
NIT−	−202,352	−166,901	−211,862	−91,019	39,565	8009	16,383	62,807
Total	6,000,032	3,155,688	10,026,464	4,875,599	7,912,302	545,399	1,189,573	1,205,401

	OIL	GAS	FBV	TXL	WOD	MIN	PET	CHM
PAD	0	0	99,835	982	179	0	10	44,100
WHT	0	0	52,730	519	95	0	5	23,292
CER	0	0	829,438	13,625	1003	0	67	204,109
CAS	0	0	1,884,568	1,191,459	55	0	12	120,668
ANH	0	3998	503,408	232,621	119	0	1	8170
FRS	0	0	61,873	4767	138,792	0	96	52,823
FSH	0	0	79,552	34	89	0	0	732
COL	0	6	22,200	28,965	541	219	0	53,243
OIL	769	130	1784	4	0	0	100,352	57,544
GAS	0	5727	2679	1178	3	0	1,354,034	5920
FBV	0	0	329,418	9020	113	0	135	44,419
TXL	0	14	51,829	2,828,257	1537	0	3	168,750
WOD	0	1	22,329	12,949	22,014	1633	2219	10,076
MIN	0	0	3600	244	67	0	963	34,051
PET	15,286	5549	46,476	56,409	2823	16,710	1733	97,057
CHM	0	101	88,521	856,042	8246	6752	54,993	2,323,537
PAP	0	33	71,814	37,186	903	88	8978	397,789
FER	0	2	9687	37,622	416	287	2447	134,411
CEM	3796	422	76	2	7	0	422	689
IRS	0	40	985	11,238	990	0	0	11,000
ALU	0	0	15,334	4539	834	0	412	75,438
OMN	17,159	5838	41,407	80,771	5458	7835	991	242,883
MCH	14,962	4447	31,706	80,422	1409	13,229	5972	46,312
HYD	905	1141	16,583	112,468	1152	5698	2998	116,469
NHY	2518	2566	52,833	358,309	3671	18,154	4127	371,058
NUC	59	80	1122	7606	78	385	13,148	7877
BIO	0	350	60,196	25,934	312	0	279	5427
WAT	0	653	3459	7541	70	1	1490	19,625
CON	4584	1881	5912	10,675	374	740	503	3722

	OIL	GAS	FBV	TXL	WOD	MIN	PET	CHM
LTR	1619	914	96,853	313,181	10,215	1781	35,648	213,131
RLY	1290	517	28,099	31,851	1664	1146	49,201	51,445
AIR	115	44	6142	19,860	648	113	2261	13,515
SEA	207	114	12,898	41,705	1360	237	4747	28,382
HLM	0	0	0	0	0	0	0	0
SER	45,702	9101	842,114	1,613,930	46,200	18,696	265,216	1,326,801
GVA	592,503	226,775	1,264,636	2,747,763	254,290	362,434	262,772	2,356,363
NIT	15,841	7790	223,189	436,660	20,612	8127	671,033	982,372
Total	717,316	278,233	6,865,285	11,216,337	526,339	464,265	2,847,278	9,653,200

	PAP	FER	CEM	IRS	ALU	OMN	MCH	HYD
PAD	83	6190	0	0	0	1466	8	0
WHT	44	3269	0	0	0	775	4	0
CER	7767	28,043	278	0	0	1806	3	0
CAS	13	2761	0	0	1	1274	9	0
ANH	338	1857	0	1	22	19,599	36	0
FRS	58,086	6855	11	1485	423	21,023	2686	0
FSH	47	365	167	0	21	17,416	33	0
COL	30,400	11,369	82,533	227,760	18,253	77,792	14,294	0
OIL	0	25,002	0	102	24	632	153	0
GAS	72	79,744	10	4151	1186	1695	4944	0
FBV	484	5405	0	104	44	618	1	0
TXL	16,272	33,724	46,624	3255	827	35,051	9034	0
WOD	1824	891	777	6736	1946	53,507	48,069	0
MIN	863	55,056	90,875	124,277	107,926	170,069	389	0
PET	7696	25,454	10,372	190,558	51,584	276,943	92,535	0
CHM	89,501	374,207	6838	38,516	56,638	434,595	278,275	0
PAP	271,317	22,253	873	2936	807	48,303	37,404	536
FER	1718	114,769	0	221	1222	7729	6584	0
CEM	3	6	26	54	1	75,583	75	11
IRS	1965	1119	6677	1,244,865	38,487	1,075,815	841,935	1249
ALU	3087	5140	723	156,054	419,868	544,457	405,766	119
OMN	17,564	29,502	25,316	418,929	21,632	1,336,044	378,590	6835
MCH	6644	9997	6502	27,903	6890	190,452	1,291,211	30,157
HYD	24,781	18,157	30,898	73,745	56,973	93,917	48,513	83,038
NHY	78,951	57,847	98,438	234,943	181,509	299,211	154,556	264,549
NUC	1676	1228	2090	4987	3853	6352	3281	5616
BIO	1654	413	2	1	2	9526	379	0
WAT	1584	5301	230	2145	1328	4953	5047	17,993
CON	1013	562	1803	6774	1130	21,990	27,248	31,300

	PAP	FER	CEM	IRS	ALU	OMN	MCH	HYD
LTR	29,069	35,889	16,108	97,579	54,153	279,768	250,524	28,803
RLY	16,681	14,187	41,704	206,023	21,433	122,063	58,159	69,076
AIR	1843	2276	1021	6188	3434	17,740	15,886	1826
SEA	3871	4779	2145	12,994	7211	37,255	33,361	3836
HLM	0	0	0	0	0	6	23,115	7
SER	156,998	219,241	118,543	683,041	224,982	1,504,300	1,401,923	147,904
GVA	163,526	395,871	203,472	755,391	366,519	3,635,882	2,746,196	570,761
NIT	84,035	145,086	22,836	382,763	115,149	814,321	802,581	69,764
Total	1,081,469	1,743,816	817,892	4,914,482	1,765,478	11,239,931	8,982,808	1,333,380

	NHY	NUC	BIO	WAT	CON	RTM	RLY	AIR
PAD	21	0	10,021	0	286	4438	0	0
WHT	11	0	5321	0	151	18,409	0	0
CER	68	0	155,354	33	317,933	107,414	0	0
CAS	223	0	3293	0	3108	0	0	0
ANH	34	0	4304	0	12,291	0	0	0
FRS	69	0	4019	0	111,191	0	169	0
FSH	1	0	88	0	83	0	0	0
COL	615,807	26,419	74	0	1781	0	5619	0
OIL	0	0	2	2267	18	0	0	0
GAS	216,335	0	4	108,804	580	0	14	0
FBV	9	0	15,716	0	346	6411	0	0
TXL	972	0	5111	447	21,613	9068	1621	9
WOD	59	0	190	63	245,724	335	1904	7
MIN	789	0	5	0	459,129	0	0	0
PET	51,536	2211	5780	7109	209,919	1,275,733	76,373	40,174
CHM	2901	4724	2881	4928	259,005	214,757	3627	32,066
PAP	1349	23	1164	674	6762	5382	1996	211
FER	82	0	3704	52	34,270	0	19	0
CEM	11	0	3	0	691,402	0	0	0
IRS	3147	53	195	952	1,238,052	106	22,977	0
ALU	300	5	46	9	22,740	30	0	0
OMN	17,219	291	5335	5806	522,099	268,547	36,002	37,725
MCH	75,973	1285	1260	5400	525,300	67,556	366,765	1501
HYD	209,191	3539	295	14,112	12,506	20,815	19,864	2171
NHY	666,458	11,274	938	46,394	39,844	66,313	63,285	6917
NUC	14,148	239	20	954	846	1408	1343	147
BIO	5	0	1663	111	3445	848	0	0
WAT	45,329	767	34	31,979	3515	2391	306	546
CON	78,851	1334	9559	21,421	39,303	42,204	107,145	15,575

	NHY	NUC	BIO	WAT	CON	RTM	RLY	AIR
LTR	72,561	1227	8797	7674	317,243	267,418	11,344	16,843
RLY	174,018	2944	2314	2397	135,602	65,988	19,959	2611
AIR	4601	78	558	489	20,117	18,721	719	323
SEA	9663	163	1171	1037	42,246	39,313	1511	679
HLM	18	0	3200	0	2081	209	102,102	5
SER	372,604	6303	51,612	50,764	1,234,596	1,097,490	105,727	41,045
GVA	1,437,874	24,324	1,048,536	222,432	4,828,003	3,541,510	1,014,773	238,028
NIT	175,752	2973	0	53,387	514,208	429,044	45,737	36,197
Total	4,247,989	90,178	1,352,570	589,696	11,877,338	7,571,856	2,010,902	472,780

	SEA	HLM	SER	PFCE	GFCE	CAC	EXP	Less IMP	Total
PAD	37,596	11,209	188,899	4,502,945	10,135	156,257	69,346	8104	6,000,032
WHT	3793	5920	99,770	2,378,316	5353	82,530	36,626	17,624	3,155,688
CER	1373	24,147	304,745	5,589,725	13,320	66,880	209,248	151,011	10,026,464
CAS	0	0	118,476	1,189,102	466	34,964	101,209	9479	4,875,599
ANH	0	10,130	294,283	5,515,106	100,496	101,258	43,970	53,877	7,912,302
FRS	0	0	35,760	108,581	99	0	8622	76,324	545,398
FSH	0	0	4496	1,067,298	373	0	3250	1240	1,189,573
COL	0	0	49,183	18,016	209	2362	5193	203,569	1,205,401
OIL	0	0	10,914	0	106	68,483	0	804,653	717,316
GAS	0	63	31,981	11,731	6559	21,579	0	226,953	278,233
FBV	611	1650	243,154	5,458,350	1085	276,919	638,976	355,000	6,865,285
TXL	153	2826	226,332	5,378,542	12,102	153,143	2,359,216	238,220	11,216,337
WOD	13	907	70,010	4369	19	−2131	11,118	6428	526,339
MIN	0	0	21,690	0	616	−41,672	116,202	690,851	464,265
PET	17,173	5967	87,645	818,973	102,082	−285,818	125,373	901,042	2,847,278
CHM	64,257	673,516	592,438	1,257,555	439,438	1,843,294	722,441	1,234,430	9,653,200
PAP	0	1833	184,922	23,510	1895	71,216	17,582	135,840	1,081,469
FER	0	44,159	30,961	44,098	12,840	48,781	23,411	234,699	1,743,816
CEM	0	0	8579	0	0	12,927	24,302	86	817,892
IRS	0	52	185,635	0	5	491,124	227,716	492,953	4,914,482
ALU	0	0	85,918	0	0	467,241	56,808	500,161	1,765,478
OMN	37,799	34,689	490,226	168,1945	349,955	420,2397	151,2176	698,335	11,240,332
MCH	3949	808	182,641	711,946	343,600	5,690,012	306,859	1,358,033	8,982,808
HYD	2284	4142	144,958	129,412	6290	0	0	0	1,333,380
NHY	7277	13,196	461,818	412,292	20,039	0	0	0	4,247,989
NUC	154	280	9804	8752	425	0	0	0	90,178
BIO	15	1080	29,942	1,095,035	0	0	0	0	1,352,570
WAT	10,153	1381	17,307	259,202	144,183	0	0	0	589,697
CON	13,873	14,767	801,355	0	698,378	9,526,650	0	0	11,877,338

	SEA	HLM	SER	PFCE	GFCE	CAC	EXP	Less IMP	Total
LTR	16,053	49,901	1,611,354	2,779,498	390,533	179,813	569,902	351,636	7,571,856
RLY	2289	8375	84,246	461,206	105,371	52,607	89,178	18,106	2,010,902
AIR	0	3164	102,180	176,254	24,765	11,402	36,139	29,666	472,780
SEA	0	6645	214,577	370,134	52,006	23,945	75,891	109,237	945,902
HLM	132	49,317	24,359	1,412,361	537,508	0	0	0	2,157,312
SER	71,970	199,901	3,700,793	14,948,266	6,872,533	1,073,078	1,521,287	314,748	40,945,178
GVA	521,701	892,354	29,569,919	0	0	0	0	0	85,717,665
NIT	133,284	94,932	623,911	1,745,903	209,587	1,212,295	0	0	9,534,000
Total	945,902	215,7312	40,945,178	59,558,423	10,462,370	25,541,536	8,912,041	9,222,304	

Bibliography

Bohm P, Russell C (1985) Comparative analysis of alternative policy instruments. In: Kneese A, Sweeney J (eds) Handbook of natural resource and energy economics, vol 1. North Holland, Amsterdam

Central Statistical Organisation (1994) Enterprise survey 1993–1994. Hotel Restaurants and Transport Sectors, Government of India

Central Statistical Organisation (2000/2004) Annual survey of industries. Factory Sectors, Government of India

Central Statistical Organisation (2008) National accounts statistics: factor incomes 1980–281–1999–2000. Ministry of statistics and Programme Implementation, Government of India

Central Statistical Organisation (Various Years) National accounts statics. Ministry of statistics and programme implementation, Government of India

Department of Agriculture (2000) Costs of cultivation studies. Government of India

Darwin R, Toll RSJ (2001) Estimates of the economic effects of sea level rise. Environ Resour Econ 19:113–129

Dell M, Jones BF, Olken BA (2008) Climate change and economic growth: evidence from the last half century, Working Paper 14132, National Bureau of Economic Research (NBER)

Devrajan S, Lewis JD, Robinson S (1991) From stylised to applied models: building multi-sector CGE models for policy analysis, Working Paper no 616, Department of Agriculture and Natural Resources, University of California, Berkeley

India's Greenhouse Gas Emission Inventory: Report of Five Modelling Studies, Ministry of Environment and Forests, 2009, New Delhi, India

India's Initial national communication with United Nations framework Convention on Climate Change (NATCOM I), Ministry of environment and forests, New Delhi, 2004

Jayadevappa R, Chhatre S (1996) Carbon emission tax and its impact on a developing country economy—a case study of India. J Energy Dev 20(2):229–246

Lin S, Cwhang TC (1996) Decompositon of CO_2, SO_2 and NO_x emissions from energy use of major economic sectors in Taiwan. Energy J 17(1):1–17

Pradhan BK, Roy PK (2003) The well being of Indian households: MIMAP—Indian survey report. Tata McGraw Hill, New Delhi

Roy J (2007) De-linking economic growth from GHG emissions through energy efficiency route-how far are we in India? The bulletin on energy efficiency

UNEP (2008) Green jobs: towards decent works in a sustainable, low carbon world. United National Environment Programme, United Nations Office, Nairobi. http://www.unep.org/civil-society/Implementation/GreenJobs/tabid/104810/Default.aspx

B. D. Pal et al., *GHG Emissions and Economic Growth,*
India Studies in Business and Economics, DOI 10.1007/978-81-322-1943-9,
© Springer India 2015

Websites

www.indiastat.com
www.nhpcindia.com
www.npcil.nic.in
www.moef.nic.in
www.mospi.nic.in
www.adb.org
www.ilo.org
www.ntpc.co.in
www.ipcc.ch